The Medical-Chemical Answer for Swine Flu & Avian-Flu

The Medical-Chemical Answer for Swine Flu & Avian-Flu

F.J. Sawaya M.D

University of Michigan. 1969

To order additional copies of this book, contact:
Xlibris Corporation
1-888-795-4274
www.Xlibris.com
Orders@Xlibris.com
67370

Contents

Dedications 7

Appetizers 11

Update 2005-2009...13

Schematic Flu Diagram 18

Avian Book Original 20

Influenza A Type Viral Test 24

Tamiflu and Betadine 27

Disclaimer..30

Coin Flip Fatal...31

Section I: Introduction...37

Section II: Meet the Press 50

Section III: Betadine™ is Povidone Iodine....................55

Section IV: Science—Layterms 68

Section V: Betadine™ Diversity. 81

Section VI: Doses and Prices 97

Section VII: Anarchy...107

Section VIII: Nutritional 113

Section IX: Expanded Uses of Betadine 117

Section X: Doses and Costs of Vitamins, Lavbac........130

Section XI: First Aid 138

Section XII: Summary...147

Section XIII: Japanese Study of Avian Flu and Povidone Iodine 170

References..173

This Book is Dedicated to:

Dorothy and Al my parents, Andrea
my daughter and her husband John
Sam Muraeky and W. Kostrubiec
Joyce Papa of Fisher Scientific
Pat and Charles Rutherford,
The World Health Leaders
Don Freeman, Ann Arbor, Michigan.
Simon Dukes, Manchester England
Marc Jackson of Texas
Gregory Johnson of Michigan

Contributors

John E. Magielski M.D. former Professor
of ENT Univeristy of Michigan
Patrick Tellez, M.D.,Ohio
Karl O Bandlien M.D. ,Transplant Surgeon
Herbert Roth M.D.,Pediatrician
Alan A Yezbick D.O. General Practitioner

Killer Viruses have made their move

The virus is bypassing normal steps and heading directly to man . . . .
a gentic mating ritual is happening joining the two viruses in 2009.

Swine & Avian-Flu – H1N1 & H5N1

A Medical Survival Guide for a lethal outbreak solely aimed at people.
Medical Contents will help you learn basics and survive a lethal
Outbreak.

A Beautiful Ending To a Sad Story
A Non-Vaccine Approach to H5N1, and others . . .

Immediately Available

The Medical-Chemical answer for
Influenza A and B is
POVIDONE-IODINE

Proven to Kill Avian Flu and Swine Flu
A forgotten medicine 50 years old!
. . . maybe this old-timer can help us out.

True incidents from around the World, 2006-2009

The **only** thing stopping a massive outbreak worldwide is a missing putter in the virus's golf bag and **only** that missing club separates you and I from dying in a human to human transmission. The Swine and avian viruses have skipped animals and **get to the green in one shot** and they land in your respiratory tract. Adaptive shoplifters is what viruses are and the *hybrid of avian and swine flu is so smart it could buy the missing gene it needs online!*

A case of Small Pox and a real case of the Bubonic Plague (which killed 1/3 of Europe in the Middle Ages) were reported in California in April of 2006.

In **1918** this truly happened: One woman who boarded the tram alive and well was dead 3 miles later! The death is from rampant, fulminant lung hemorrhage.

Most viruses are one/ one thousandth the size of a pen point or **1/40th the diameter of a hair on your head!** The enemy is fast and small, penetrates most masks or barriers like a 22 bullet!

The **Tamiflu** is *not* designed for human to human transmission.

Certain methods discussed here are theoretical and need permission from your doctor or the CDC to implement, but enjoy a real way to save you and your family's lives.

Japan this April of 2006 published proof that Povidone Iodine is the generic name for Betadine™, **kills Avian Influenza A.**

In **Thailand** recently was a story of a child who ate an infected chicken, and **died within 3 hours of hospital arrival**. The same girl had also infected 3 other women by airborne vectors.

Avian Flu. Its size—very small predisposes to rapid spread and it is **only 1/40th the diameter of a human hair.**

How this Vietnam Infected Family . . . coped with disaster is detailed in the book. The Immune father may never have *been antibody screened!* or at least never revealed . . . and he was never sick so it's possible that he can *still be blood tested THIS 2009* and save the scientists some immunity dilemmas, and deaths. November 2005.

Y2K – is a perfect example of the world preparing for a computer failure, saving food, and water; whereas this Avian Flu, SARS business is far worse than any Y2K could ever have been., and very few people save anything at all!

The *new principles* in the book are applicable to any and all airborne virus not just the Avian flu.

Update covering Avian Flu Outbreak of 2005 to Swine Flu of 2009.

June 2009, about 1000 USA cases Swine flu and about 50 deaths
In the 100years..prior to that it is estimated only 100 people died from Swine Flu.

Fact of Life is on the line.
The principal simply is that an infected bird can kill you and an infected pig can kill you, so we have to stop this and you as a reader must accept the fact you know less because some do not want you to know what ails you!

The thing stopping a massive illness is a missing 9 iron in the virus's golf bag One club separates you and I from a human to human transmission and a virus Is so smart it could buy the missing gene it needs online!

Very little vaccine can be manufactured to solve either of the two problems.So, read and learn and open your mouth when possible to get some useable help, as you will see reading more.

10,000 Jamaicans were swiftly killed about 10 yrs ago when a ship from Bangladesh blew ballast into the harbor and the ballast was cholera! The city closed the people all died . . .

VHSv virus killed 100,000 TONS of Michigan Lake fish
Swiftly One Hundred Thousand Tons gone, this was Several years ago but the virus is back keeping the Mercury poisoning company. VHSv came from somewhere?

Population Density and Respiratory Receptor Cells

Population Density means crowded places. so by understanding that the Human Respiratory tract is now open season to H5N Avian Flu and the new mutating Swine Flu,H1N1, the closer living quarters are Lethal and nothing less. Overcrowding opens infection routes.

This means that the viruses of the bird or pig can kill a few billion of us in about 6 weeks if humans start infecting humans by coughing and sneezing. *The Influenza A type virus can get into us without a intermediary system There are not enough gloves or masks in the USA to even protect 10% of the population*

Overcrowding is the main ingredient in the likelihood of a pig infection beginning, and is less likely in smaller groups o than in larger groups. This epidemiology is true for human colds a lot of kids in a classroom will make more colds than a bunch of smaller classrooms and the reason the H1N1 of 1918 would eradicate 1000 x more people now is we are like sardines now compared to 1918. No planes and no overcrowding.

Then there were no planes no high rises, and It killed 100 million internationally and 30 million USA was the often found term It is a well hidden fact never taught in medical schools. Somewhere between 30 million and 100 million died in less than 6 months. And I will support this accuracy with a personal experience. An unnamed man from Columbus

Area in Ohio, ***mentioned his great grandmother told him that In 1918-20 about, that 7 of every 8 houses on the block Were filled with dead people! –Occupants of SEVEN of TEN homes/ block*** **In Ohio were killed by H1N1.**

This I believe since the man is very very prominent and I also Know this is a hidden untold secret of mankind now going on 90 plus years of " Don't tell" it is the Roswell Park of Public Health.

The *population density* of people and of pigs and birds are the factors to consider. Imagine if an Asian plane lands here in New York, and some passengers are filled with Sars Virus or Avian-Swine-Hybrids, and the passengers switch planes for Newark, Nevada and Seattle all within the hour! The follows t-he high speed seeding of highly populated areas will bring mankind down.

It is the main factor for rapid spread of any disease. The virus from getting into the bird and into the pig or from the pig and bird to humans, and worse from airborne human to human transmission of killer viruses swine, swine hybrid or avian hybrid.

Purpose:
Teach Influenza A is Swine Flu and Avian flu, They are the Hatfields and the McCoys. This book describes a technique which will kill all microbes on earth and remain human safe for air and water. You will read genetic truths, unspoken dangers to your life, and a proven way to kill the swine, and the avian flu and more with an old fashioned medical chemical.

We have 6 billion people on earth ,43 billion chickens earth and 60-millon pigs! This allows a high-degree of genetic intermingling of people and viruses and adaptive genetic changes harmful to mankind.

Since Avian and Swine flu are familial in habits and both attack humans directly and indirectly. The facts in this book are safely said as being the same habits or similaties for both the Swine flu.and Avian flu.

The book has the methods of quarantine, identification of infected..animals and humans. vitamins and herbs used for preventitive medicine.

Both viruses are almost twins. So you can safely insert the swine for avian words When it comes to mutations living places, hosts,

The principal simply is an infected bird can kill you
An infected pig can kill you, so we have to stop The virus from getting to the bird and pig or from The pig and bird to humans, and worse from animals to an airborne human to human transmission of killer viruses.

They have all begun to become identical twins since the book was written. The nice thing for the reader is the treatment suggested is part of a whole new way to clean our air and water.

A *Proven Method* *of destroying these viruses* in confined land areas or water areas is discussed in the book and most recently on CNN in June last year was this: PVPI or Betadine™ was inserted successfully into the breast-milk pump of HIV breast-feeding mothers and the PVPI, Betadine™] successfully stopped HIV transmission to the babies and with no harm.

In 2006 Japan published verification of my theory of Betadine™ to kill Avian flu, but better yet,the recent HIV news is great for the Reader to have down to earth scientific prophecy come true. Since the first writing many unproven items of PVPI which is the chemical of Betadine™ came true. It can cure many things and kill every pathogen if the dose and timing is correct. PVPI is the generic form of the orginal Betadine™ patented 50 years ago.

Genetics and virologists show that the margin of difference between Avian flu and Swine flu is not as wide as before and is getting smaller all the time. The book here is geared to Killing Influenza A type and references Avian Flu most often . The Avian and Swine viruses are the same family carry the same guns just one a Winchester the other a Browning both are guns and do similar things.

Trust my training and references, that what is good for one is good or true for the other virus of concern.

The book teaches. Biology, Chemistry,VirusTypes, Holistic Prevention, Medical Prevention.

Lethal Gene Stealing occurring.
When the avian flu hits the same turkey infected with swine flu we have a genetic X-file-Tupperware party open house! If Avian Flu learns to triple adapt like the Swine flu it will kill billions of us much easier than the Swine flu, due to a simple fact that there are more birds than pigs,and birds can fly and pigs don't fly . Human to human transmission is a capability proven in

Asia with horrible deaths. Once in place the Virus cares not who it kills, but both the Swine and Avian flu have learned to go directly to humans. Skipping the " reorganizing stages in animals. The respiratory receptor cells in humans make it possible for Swine and Avian flu to kill us. Until recently the viruses had to take " the long way home" to people, now they take-have a shortcut.

Everytime the Influenza A team members,Swine or Avian flu go through.. the hosts..Birds turkey pigs people they have a chance of grabbing onto a **human gene** and of making it into a. Primary Human Transmitter* and killing billions if it happens.

WHO Director-General Dr Margaret Chan.

International Human Cases of H1N1
Influenza pandemic alert raised to phase 6
11 June 2009 — On the basis of available evidence and expert assessments of the evidence, the scientific criteria for an influenza pandemic have been met. The Director-General of WHO has therefore decided to raise the level of influenza pandemic alert from phase 5 to phase 6. "The world is now at the start of the 2009 influenza pandemics

Read Dr Chan's statement to the press
world health organization total(cases), & deaths to date june 12, 2009 H1N1

WHO | Influenza A(H1N1)—update 46
10 *June 2009* — As of 06:00 GMT, 10 *June 2009*, 74 countries have officially .. Laboratory-confirmed *cases* of new influenza A(*H1N1*) as

17

officially reported to WHO by States Parties to the International *Health* Regulations (2005) . . . Venezuela, *12*, 0, 8, 0. Viet Nam, 15, 0, 6, 0. **Grand *Total*, 27737,** 141, 2449, 2 . . .

On or about June of 2009 25,000 world cases of swine flu had been reported but over 17,000 were in the USA and most between Texas and Chicago,Illinois the old meat market route,to the slaughter houses of the past. There were 45 deaths up to June 24th, 2009 estimated.

The USA seems to have accounted for almost 2/3rds of the cases up to June, 2009/

A SIMPLE WAY TO SEE VIRUSES MOVE IN ON NATURE

Avian Flu Sequence
A (Original pathways)
Virus→ ducks (Asia)→ , uncooked duck→ eaten it infects humans-Humans→ get better or→ die after eating partially cooked infected ducks.(Ducks wild birds turkeys) ⇔ **human →human = pandemic., the pandemic has not occurred yet.**

The denial of human to human transmission is refuted in the book It occurs enough to make the smart scientists pray.

B (New Pathways)
Newest progression since 2005
Virus→ ducks →, pigs <-> humans ⇔ human →human = pandemic

The denial of human to human transmission is refuted in the book It occurs enough to make the smart scientists pray

Strong mutation capacity makes a sneeze from a pig onto me a lethal weapon which would get me to church or prayer quickly! The petting the Pig at a farm was another reported severe Case of Swine Flu this past June 2009.

Swine Flu Sequence

Ducks → turkeys → humans <-> pigs

Virus→in ducks → poultry → humans ⇔ pigs.

The pig-bird-and human 3 : 1 version was in.a farm worker in Canada in 2006.

If a bird pops up in a swine pig pen which is a carrier of the avian strain it can fly off with a horrible mutation.

The Swine flu is a triple mutant which you might call land sea and air. It is in soil around the farm, it is in ducks and birds around the farm and in the water also.

And when it bumps into brother Avian inside a sick Turkey we have a turkey-shoot for a mutation a joining or mating of the two nasties of Swine & Avian . Lets call it Swavian, for short since they are much like gentic *non-identical twins* for a new way to think.

H3N2 virus was discovered in pigs in North Carolina in 1998 and it looks like the same one that pestered humans since 1968.

The virus went cross-country in one year. Foreign people such as Denmark limited the number of pigs allowed together, but USA did not do the same. The US was too liberal compared to Denmark.

Argument with back up—LETHAL IF NOT COOKED

Lethal if Not Cooked Well is a Poster-sign you might see if folks are not truthful with us and open and careful. Yes the ***Lethal If Improperly Cooked*** is Historically true and documentable, and terrifying. All Asian reports.

Much has been denied about " you cannot get sick eating avian infected birds. which is a lie and there are documents in the book of mass graves in Asia from family cookouts, and family dinners. As far as swine flu

19

being upper respiratory, in view of the mutation capacity, a sneeze from a pig onto me would get me to church fast, but it would make me really, *really incredibly nervous* to eat pork during a Swine Flu outbreak...*since all genetic rules* sort of end lately with pig improvement and humans worse off! And with a hip-hoping gene stealing virus skipping from pigs to birds and back and forth humans to pigs and why say more.

Some of many infected folks died rapid transmission respiratory deaths like the Movie Outbreak with Dustan Hoffman. Why then would anyone encourage people to eat pork—when the *Avian infected duck-if eaten will kill you* if partially cooked and since June maybe the Avian virus will just skip dinner and go directly to you. Avian and Swine are spelled differently but are becomingTwin-hybrids aimed at you and me, not birds or pigs! How do you or anyone on earth know when the mixing of the avian in the pig becomes **lethal if not cooked fully. So you be the judge.**

Likely it will be a farm full of pigs in slop who come down with the hybrid-triple adaptive gene, not just one pig and a side of bacon for sale at your local store.

The story does not stop here H5N1—Avian Flu can jump from Human to Human and I have included a few examples much like the "Outbreak Movie" viruses, avian strain just horrible human to human cases in Thailand –Asia. The book focuses on the Avian Flu of 2005 and how a well-known medical human-safe chemical can **kill living Influenza A organisms.**

Avian flu + Swine flu = almost Twins

Since Avian and Swine flu are familial in habits and both attack humans directly and indirectly,the facts in this book same for both the Swine flu. and Avian flu.

The methods of quarantine,identification of these viruses in infected.. animals and humans and preventitive medicine for both are nearly identical. So, you can safely insert the swine for avian words—interchangeably as far as host pathways and severity. They are as of June 2009 becoming a hybrid virus anyway!

Pigs cannot fly but a sneeze travels 20 feet and Pork is the most popular meat around some places.

When it comes to mutations living places, hosts They have all begun to become identical twins since The book was written. The nice thing for the reader is the **treatment suggested is part of a whole new way to clean our air and water.**

Birds Pigs and Your Drinking Water:
Avian flu scares from 2005 in Asia and the world are viruses transmitted by air/ and water, and from ½ cooked ducks in Thailand.. People died hideous deaths. Story in the book .

Water Transmission of lethal viruses
is occurring and proven

There is proof of If influenza A being located in a pond and infecting normal birds, (which makes one wonder about The water supply)

#1

*www.who.int/water_sanitation_health/emerging/h5n1background.pdf
—Similar pages*

#2
and also from
Water-Borne Transmission of Influenza A Viruses?
Jul 23, 2008 . . . Key Words. Influenza A viruses; Ducks; Transmission; Water . . . *The isolation of influenza A viruses from unconcentrated lake water and from . . . content.karger.com/ProdukteDB/produkte. asp?Doi=149014—Similar pages* by VS Hinshaw—1979—*Cited by 52—Related articles—All 5 versions*

Reference Sited: this page only.
birdflubook.com/a.php?id=58&t=p—18k—Cached—Similar

Death report: bird virus and human virus was in the pig virus

This reference which I read I chose to recommend to the reader as it is deep and scholarly and upbeat for genetics and virology and it does document the point of my book which is *both viruses may be working together to kill people.*

The world's bans on Asian poultry because of bird flu, combined with bans on U.S. beef because of mad cow disease

Reference below Bird Flu:

Below:::

Since **pigs** display both bird-**type** and human-**type virus** receptors, . . . CDC laboratory analysis showed she was killed by a **swine flu virus** that she . . . **waterborne** intestinal duck **virus** and a killer **airborne** respiratory human **virus** . . .

birdflubook.com/a.php?id=58&t=p—18k—Cached—Similar

(Waterfowl excrete influenza viruses into water)

The Book As Written in 2005-2006 and Updated a bit for Swine Flu but both Avian and Swine are nearly *Identical* in actions. April 26, 2006

News Breaking Bulletins: 2006

.U.S. efforts won't slow pandemic flu if it hits scientists said on Wednesday (April 26,2006) there was no "magic bullet" to control a human influenza pandemic but a combination of measures could cut the number of cases.
http://www.msnbc.msn.com/id/12497021/from/ET/

.Outbreak of highly pathogenic avian influenza in Japan and anti-influenza virus activity of povidone-iodine products.
Dermatology. 2006;212 Suppl 1:115-8.
Ito H, Ito T, Hikida M, Yashiro J, Otsuka A, Kida H, Otsuki K.
Department of Veterinary Public Health, Faculty of Agriculture, Tottori University, Tottori 680-8553, Japan.

.A case of Small Pox and a real case of the Bubonic Plague (which killed 1/3 of Europe in the Middle Ages) were reported in California in April of 2006.

Instant Message
An instant understanding of the danger is to know that the thousands of infected birds flying around are like miniature jet airliners and are more fatal than the planes of 911. Even the non-infected bird becomes a potential aggressor. There are immunologically thousands of 911 planes flying around.

The *new principles* in the book are applicable to any and all airborne virus not just the Avian flu.

Avian Flu Book Text

*A Saving Combination exists now office tests for Influenza A
And one called the Qdel test.*

*INFLUENZA A OFFICE TEST
BETADINE™-POVIDONE-IODINE,
LAVBAC+FM (Letters representing things such as licorice anise Vitamins
,Betadine™ etc.*

Early diagnosis with the QDEL test and the Medical—Chemical use of Povidone Iodine,Betadine™ will save many, many lives.

Use the Influ A. test to screen patients, and the Betadine™ to contain an outbreak. QDELand other brand names test for Influenza A and the Avian flu is a subtype of influenza A. Swine flu is also Influenza A.

To fight a killer virus using only vaccine is unwise given the existence of a proven medical chemical 50 years old. The medication is safe for use on the brain,the eye,the ear, the nose and throat and vaginal areas of humans and animals worldwide and was the agent of choice in killing HIV in operating rooms.

The fact that Betadine™ also called Povidone Iodine (PI) has proven effects in the killing effect on airborne viruses and **should force the health people to 1)test and 2) create another synthetic antivral chemical medication able to be used like Betadine™.**

The use of an airborne human safe virus killer used 50 years ago would certainly lead the reader and health leaders in a safer direction than currently provided.

Nothing has ever been created as strong as Betadine™. It has a broad antimicrobial range and is proven to be human-animal safe for almost half a century. The absence of anything close to its medical effectiveness I find odd, repairable, and needed.

The Two Best Ideas of 2006 for Viruses:

Influenza A Office screening tests such as the QDEL test and Betadine™ antiviral, antimicrobial and 100% efficient depending on dose and time coverage and also called Polyvinlypyrrolidone Iodine

1) The Office Influenza A—Screening Test
   (Ex. QDEL Screen Test For Influenza A and B.
2) The *Chemical Medical Containment of Avian Flu.*

The Screen Test: Two or more tests are available to the doctor office to help diagnose your family.

There exists a test for Influenza A and B which is called the QDEL test which takes 10 minutes to do in a doctors office and can effectively screen the entire USA for routine flu including the subtype of the Avian flu possibilities.

This test if coupled with a chemical medical approach to prevention and containment of viruses will help lock out the airborne viruses forever.

Both are needed to screen and prevent a pandemic. The QDEL test takes 10 minutes in a doctors office, and about 30% of Michigan physicians can perform it while you wait.

Avian flu is a subtype of Influenza A, and here is how the test will save your life. Avian flu must be treated within 48 hours of diagnosis or the infected person becomes part of the 50-72% mortality rate sector. Tests for Influenza A will take 10 minutes now! So a working diagnosis is instantaneous!

The instant read out lets a person start on a dual treatment including influenza A and a possible Avian flu—so the doctor and patient are happier. Tests for Avian flu take 72 hours or better; so if one is positive for Influenza A by the QDEL test then the patient can be treated for Influenza A as well as for *a presumptive case of Avian flu* until the 3-4 day Avian test results gets to the doctors office. Instead of missing Tamiflu for the 48 hour Avian critical treatment period, the Positive Influenza A patient begins getting immediate Tamiflu care and has received perfect protection for both Influenza A and the Avian Flu.

If this QDEL test is positive for your child or adult Then Tamiflu is given automatically and very effective. It also begins treatment for a "Maybe Case" of Avian Flu until the Avian test comes back.

The fatal gap of time to test for avian flu/and now possibly Swine flu of 72 hours is filled by the 10 minute QDEL test and it will for sure save lives without doubt,guesswork or hesitation. If the Avian test is positive then one avoids lethal complications. *Remember in order to save your life the treatment of Avian Flu must be started within 48 hours of SYMPTOMS, and the QDEL test allows treatment within 10 minutes of entering the doctors at a cost of about $ 25.00 only, not 72 hours!*

Influenza

Diagnostic test provider obtains U.S. FDA clearance for influenza A + B test
January 24th, 2006
Quidel Corporation (QDEL) reported that it has obtained U.S. Food &
Drug Administration (FDA) clearance for its claims for the
**QuickVue Influenza A+B test, which is a 10-minute diagnostic test
for influenza.**

Among new claims added to the package insert is 94% sensitivity for
detecting influenza A when using nasal swab specimens. The package
insert for the test is being updated to include clinical study results for
nasal swab, nasopharyngeal swab and nasal wash in addition to analytical
study results for the detection of cultured avian influenza viruses.
The QuickVue Influenza A+B test has been shown to detect cultured
avian influenza viruses, including avian . . .

1) The Chemical Medical Prevention of H5N1

.Betadine™ has been shown to kill airborne viruses in an extensive study
from 1997 and is used by practicing doctors every day for eye surgery,
herpes, and sinusitis and more. Japan has recently proven that PI-Povidone
Iodine will kill Avian Influenza A. In April of 2006 and recently the same
publishing Japanese group in June 2008 reports surfaced on CNN that
Pvpi was used in breast pumps for HIV breastfeeding mothers and it
prevented the spread of HIV to the babies of the mothers with HIV!

ANNOUNCEMENT OF A MEDICAL –CHEMICAL PREVENTIVE MEASURE AGAINST AIRBORNE VIRUSES A WORLD SYSTEM

There is an all-encompassing plan in this book which makes use of
Betadine™ in and on the human body, in all facets of human life globally.
The plan allows protection of all humans and animals at a fraction of
the cost of vaccine, and is broader in scope and availability. Some of the
methods are doable now, while others will depend on CDC studies.

If you consult your doctor for the QDEL Test or simlar other brands and follow the concepts in this book, you will lead a safe more educated life in the viral age of the 2000's.

The lifespan of Tamiflu is only one year of shelf life so the need to use it in conjunction with the QDEL Influenza A test is critical for any person with flu symptoms. Also remember that about a billion doses are around but will expire in one year or less and **the Tamiflu is *not* designed for a human to human transmission.**

Also, since there is **NO VACCINE for human to human transmission in exisitence,** it is important to have an alternate plan to save our lives whether a killer mutation comes this year or 10 years from now; sooner or later a mutation will occur,and for a world to be waiting unprotected is almost fatalistic. If a country had a nuclear threat aimed at us and others and capable of terminating 3 billion lives, much time would be focused on the problem. **The virus has this much power and needs our attention.**

The book is a "cookbook" on how to protect your family home and office building, with medical aids, dietary and theoretical and also implements a proposed filter and aerosol system. These aids provide you some **structure for survival** in the absence of vaccine or in conjunction with a useable vaccine.

It is a 'Joe Public' plan for the World for humans,animal and can be easily implemented and is affordable to any or no income bracket. The Povidone Iodine concept is supported by pages and pages of laboratory research from 1997 Povidone Iodine studies and 2006 the month of April. THE FOREIGN STUDIES are ignored by our laboratories. and will be needing some testing and verification by the CDC and other concerned, able parties.

And by adding some 3000 year old ancient nutritional remedies to the medical uses of the best humanly safe airborne-virus-killer Betadine™ we may be able to offer **an immediate preventive shield** against Avian Flu, and possibly Ebola, SARS and Small Pox in the presence or absence of vaccines. The shield will cover normal every day life and agriculture applications as well. **There is no existing medical-chemical or mask system** for sale on the table to back-up the deficiency in vaccines except for the systems proposed in this book.

In collaboration with another medical doctor, a medical immunologist and other people in the medical fields, we have devised a theory that merits the support of further study by the CDC on the efficacy of Betadine™ on airborne viruses H5N1, Ebola etc.

Betadine™, which is as readily available as water, and **cheaper** than paint by the gallon! $ 22.oo retail and the Povidone Iodine about is $ 15.oo /gallon and available for much less.

F. J. Sawaya M.D.
University of Michigan Medical School 1969

[Disclaimer]

The use of Betadine™ Povidone Iodine, is based on actual laboratory studies of 1997;and the proof of killing Avian flu Is published here from Japan April of 2006. However its effect on Ebola are not documented, but sparcely mentioned On the internet but not a Harvard type paper and the idea is still being studied at a couple major USA research centers.

A section at the end of this book has recommended doses, costs, and how to use the combinations listed in the book but of the many uses over the past 50 years, some disclosed here are accepted in medicine and the others are theory and proposals for the CDC to test. Consequently the use of Betadine™ is only to be done with supervision of a doctors advice until the health people test airborne viruses against Betadine™. And no problem should come from seeing your doctor for a QDEL test if flu symptoms and a fever of 102-103 hits you or your family.

I have **no professional connection** to the CDC.

Japan this April of 2006 published proof that Povidone Iodine which is the other name for Betadine™, kills Avian Influenza A. This is a landmark of progress since it is the first publicized proof of a non-vaccine, human-safe way to kill airborne viruses

A section at the end of this book has recommended doses, costs, and how to use the combinations listed in the book but

Our Purpose for Writing the Book

Is to provide affordable Influenza A testing such as **QDEL** screening of Influenza A and B and protection worldwide against H5N1 and H1N1 Swine flu other airborne viruses for men, women and animals using **known medical and nutritional compositions** and a **new safer face barrier system** and for the entire planet. The medical and nutritional elements here are readily available on the open market, and can be readily tested for verification by the CDC,WHO for validity of performance

FATAL FLIP OF THE COIN

The Equation of Life is Upside Down for the first time in 50 years!

Warning: For 50 years medicine has conquered germs and has prolonged life over Illness, but with the SARS, EBOLA, and H5N1, but now the equation is dangerously reversed.

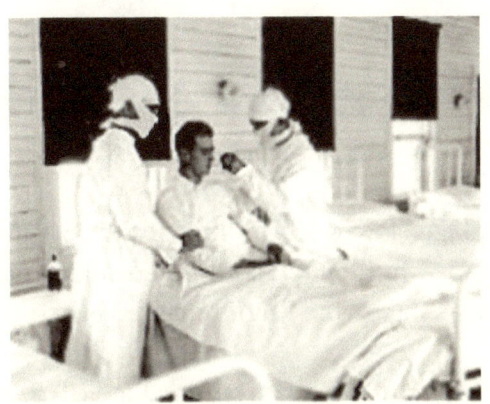

USA Army Hospital 1918

TOP of the COIN
V
1918   →   1940-2000   →   2006

MOTHER NATURE   >   MEDICINE   >   MOTHER NATURE

///
/////////////////////////////////// **LIFE'S COIN** ////////////////////////////////////
///

MEDICINE   >   MOTHER NATURE   >   MEDICINE
^
BOTTOM of the COIN

Medicine and vaccine changed the past of 1918; however; the past is back again confusing vaccines and medicines.

U.S. Army hospital 1918.

In 1918 a global plague killed 100 million people, 30 million in the USA. Reports range from ½ million to 30 million deaths. The virus of 1918 is the same notation as the Swine flu of today H1N1 so

When this book began was 5 years before June of 2009 and the deaths from Swine flu H1N1 so the book ends up being prophetically accurate and even trustworthy for advice.

However, **28% of the unlisted population** was infected in a few short months. There is a chance for annihilation of the human race due to POPULATION DENSITY and MODES OF TRAVEL.

A Recent autopsy of a man, who died in 1918, shows that he was killed by H1N1, an older flu strain; the principle remains that the 100 million in 1918 is about a billion or more deaths in 2009. In 1918 this truly happened: One woman who boarded the tram alive and well was dead 3 miles later! The death is from rampant, fulminant lung hemorrhage.

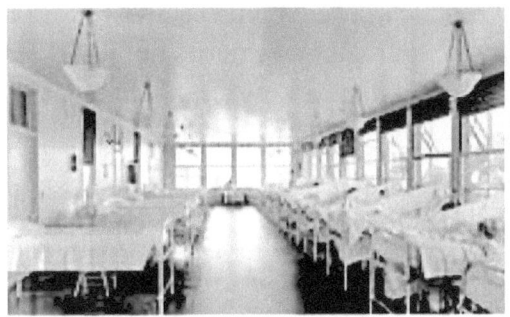

Hospital Ward 1918.

"Don't Spit" is the Sign at this Army Facility, in 1918. and indicates the fear of the times.

April, 2006.

Should you fret about bird flu? Experts weigh in
With bird flu marching across Asia and Europe and expected to reach the United States soon, scientists fear the H5N1 strain of the virus could evolve into a form that would spread more easily among humans. This has sparked fears, concerns and, most of all, questions.

http://www.msnbc.msn.com/id/12372921/from/ET/

Betadine or Povidone Iodine will kill Avian Flu H5N1 proven in 2006 Dermatology Review.www.pubmed.gov. Japan has tested Povidone Iodine on Avian Flu and Povidone Iodine will kill Avian flu virus, and needs more studies from this country's health care centers.

This insert is to remove any doubt as to the direction of *thought of this book and will make the reader a believer* in a medical chemical answer to viruses as doable. Others can just accept the fact that proof exists.

Abstract taken from April 2006, Japan Avian Flu Studies
Department of Veterinary Public Health, Faculty of Agriculture, Tottori University, Tottori 680-8553, Japan.

These results indicate that PVP-I products have virucidal activity against avian influenza A viruses. Therefore, the PVP-I products are useful in the prevention and control of human infection by avian influenza A viruses.

PMID: 16490988 [PubMed—in process]
    The above comment is the conclusion from the article listed in the Dermatology Abstract of April of 2006.

    The entire article from Japan is located toward the end of the book.

Avian-Flu – H5N1

A Survival Guide

What Everyone needs to know and do NOW in case of an Outbreak!!!!

Vaccine comes from eggs, and no one can really make Instant vaccine, actually from the onset of a killer mutant Strain to a useable supply of anti-killer vaccine can be up to 6 months, and by then many will be dead.

Hard-Boiled Reality

THE WORLD is RUNNING OUT of EGGS for VACCINE; there will not be enough uninfected Eggs for an Outbreak Vaccine!

If an outbreak occurs, 1/3 of the work force is lost in a 2 months time period, and more work force shortly after the first few months **.A Shut-Down of Global-Economy will possibly occur if an Asian Outbreak occurs, paralyzing the Chinese East-West trade balance.** (World News 3-20-06,4-26-06) Since this note, June 2009 came and went and scared people with the Swine flu, which is becoming the twin of Avian flu.

Betadine or PVPI is one of the most exiting medications in history. It exists as a Medical/Chemical—in the World to back-up the deficiency in Vaccines except for the systems proposed in this book. It is used for Herpes vaginalis, sinusitis, adenovirus of the Brain and other abcesses and even a rough-tasting mouthwash.

SECTION I

INTRODUCTION

In 2009, the facts on numbers of eggs available in pandemic is just as low as in 2006.

The following facts will change your way of thinking to a new direction.

On February 20th, 2006 National Geographic Television Special Documentary-Sponsored by Web-MD, televised "The World's Deadliest Viruses" and it was a special follow-up of the November 2005 show on H5N, Avian Flu. The same TV program stated the H5N1 is so rampant in chickens, and infects the eggs such that there may not be enough eggs to incubate vaccine.

The man who genetically clipped the H5N1 chromosome tip, and mated it with another flu, really did create a non-fatal Avian Flu vaccine; however, he stated that there may not be more than 450 million doses available, which is a tiny fraction of the need of the 6.2 BILLION total people living on earth.

What is worse is that 6 months will pass before the actual proper vaccine for the "killer" virus can be made. During that period there "**will not be**

enough eggs in the world to create a vaccine for more than 1/12th required for all people and total Global Protection!

A virus is like a Rubiks cube it will change its own shape at will. It has the ability to mutate, change and mate its DNA with other viruses; Hence, H5N1 flips its Rubiks cube structure, which it has done and **Avian flu has mutated 13 times in the last 13 months!**

How can you protect yourself with or without a vaccine?
A remedy from 3000 yrs ago, discussed in the British news papers October, 2005 was Licorice and is reported to have dramatic antiviral properties. Anise is the main ingredient of The Roche Company's Tamiflu™. Anise and licorice and pine needles are the strongest antiviral natural growing entities, St. Johns Wart and others are also good. See the nutrional section later in the book.

Reality Items to accept now.

Avian flu has in 2006 killed a full grown tiger and another large jungle animal.

1/28th ounce of infected bird fecal mater has have enough power to kill one million other birds

3 Billion people can die if the H5N1 mutates to humans.

An active case in China can board a plane and within 24 hrs and a lady in her kitchen in northern Canada can become secondarily infected by a China companion traveler contact and die within a few days!

The only recent success of making a killer safe vaccine, for **Ebola** was at a U.S. Army research ICU base in the central USA. The army made the creation of a" monkey—safe vaccine" for monkeys only! Since 2006 a vaccine for Ebola Has been made but the distribution and net efficacy are Unknown to be, and not really published very well.

The Ebola was piggybacked into the monkey on the back of a normal cold or flu virus, "snuck " its way into the monkey's immune system, making it resistant to Ebola. Ebola is the most hideous virus on earth, when infected the entire body hemorrhages internally and externally from every orifice, until the person becomes literally a soggy bloodless mess. It exists in Africa and Cambodia.

WAR: Military Says Terrorists Might Use Bird Flu as Bioweapon

On: Fri March, 11 2005 @ 03:16 GMT
On the heels of the World Health Organization (WHO) advisory for countries to stockpile H5N1 bird flu vaccine—a policy designed to spur mass-production—military reports say the bird flu virus could be used as Bioweapon

No chicken and no eggs and then what?
Around the Corner stands Blind-Sided Danger.
and a straw to break the camel's back.

Consider the danger of seeing who is at your front door
while the thief is at the back door.

In gearing up to fight the Avian flu we have found out:
that there is no chance for further vaccine after 450 million doses and that we are running out of eggs and we are unprepared for anything new or old. One more virus going wild will be the straw that can break the camels back.

The set up for a world economic chaos is possible and must be recognized, that the survivor of the present danger may not be the most powerful, but the least affected or the best protected and prepared or sadly, some entity able to inflict a viral attack which has no defense while we have no personnel healthy enough to use the vaccine delivery system and no means of making a new vaccine from lack of eggs. **If the economy is reduced 33—40%, who is around to run the vaccine dispensing shop?**

Today there are no travel restrictions on China travel, none have been enacted. There are no border precautions at the American Mexican border either.

"Outbreak" the Dustin Hoffman movie about a virus—just like H5N1, was carried by a monkey which killed off part of a small town, and the virus depicted is like H5N1 and Ebola combined.

Currently the greatest fear of scientists worldwide
Is that H5N1 or similar viruses will mutate so rapidly that by the time a vaccine arrives; 2 billion people will die in less than a 6-12 month period! This is no exaggeration! There are 6.2 billion people on earth and well over 100 billion animals. So we should take this seriously.

Stories included here are fictional and true.
The National geographic TV show in November of 2005. describes a story of "lady zero" plague starter. It is an incredible fictional story by scientists on the end of humanity. This is a story of a woman carrier who ends civilization quickly!

CNN, TV broadcast, November 2005 aired a true story "Vietnam "cluster" story of the spread of Avian Flu in an entire family! All were near death and hospitalized. The broadcast went quite unnoticed due to the late hour and never was mentioned on any local news the following day.

The story talked about H5N1 and 3 human transmissions from a cooked Asian chicken. It told of the Vietnamese family who ate a cooked chicken became critically ill.

See the details of this story later in the book.

In Thailand recently was a story of a child who ate an infected chicken, and died within 3 hours of hospital arrival. The same girl had also infected 3 other women by airborne or droplet contact . . . . only one woman lived. The child who died was 11 years old.

Fact

99-100% of all viral flu outbreaks in the past 100 years came from China Dekan peninsula where ducks and other wild animals cohabitate and breed unusual vicious strains of viral infections which spread rapidly to humans. There has never been an attempt to quarantine the peninsula, but media have recently discussed the potential of a quarantine of Thailand itself if the virus gets out of control there . There is a 2 week window to secure a quarantine from the first signs of an outbreak; plans are being formatted in Thailand now, but what is the exemption or sacredness of China's Dekan peninsula not to fall under health guidelines that affect the world? Today there are no travel restrictions on China travel; none have been enacted.

Also included in the book is a discussion of a known and accepted 50 year old medication used worldwide, which is the most virucidal human-safe agent on earth—Betadine™

It is a 50 year old medical treatment still used today by doctors and nurses and nearly everyone.

In addition known 3000 year old remedies are discussed that are used for viral and non-viral illnesses namely, Anise, Licorice and Birch bark.

Enclosed are my own theories and proposals for localized or transglobal atmospheric—air modification systems. Use Betadine™ aerosol for home and work and theoretical crop-dusting or cloud seeding for protection of larger areas by causing a viracidal rain. Traditional licorice,anise and Betadine™ aerosol,cloud seeding,filters and masks are abbreviated with the initials LAVBAC-FM(filter,mask)

An unchecked airborne human H5N1 virus has the potential to cover all continents of the earth, killing wildly in less than an eight week time period!

The only hope is early confinement and quarantine. An example is the mini-outbreak in Thailand, It was suggested that quarantine-total city and country closure would be needed if the Avian virus gets out of control in Thailand.

What is Avian Flu?

Avian Flu is "Flu like the colds we catch, but far advanced and able to kill millions of birds and even humans. It, unlike our "Cold", can be caught by eating infected chickens or ducks, and pork.

It also has the ability to be spread between humans and has been very under publicized regarding the proven cases which are discussed in this book. **Its size—very small predisposes to rapid spread and it is only 1/40th the diameter of a human hair.**

This book is to teach every individual a method of protecting themselves – families and businesses within medical guidelines. Most of the recommendations here, even medical ones, require no prescription to "Lock-Down" your house or office. So by using an old medication-chemical, and diet supplements about 3000 years old you may improve your survival chances. The use of Betadine™ has broad respiratory applications. April of 2006 studies from Japan show avian flu being killed by Povidone Iodine.

Allergies to Iodine and shellfish are prohibitive in use of one of the Betadine™, but I believe conceptually the CDC and others, will come up with a close alternative to my suggestions. Someone should create a synthetic chemical for an aerosol for human use, besides addressing the vaccine approach.

I hope you and your doctor will be able to benefit and that you will urge the health entities to research the ideas contained here. But do not use suggestions in this book without your doctor's advice!

There is no plan for an Outbreak that is either medical or preventative for the actual killing of airborne viruses except for the method described in this book. Betadine Povidone-Iodine, Chemical-Medicine can save these people, used with a mask, filter and alone.in homes, buildings offices, *but not in outdoors*.

An Avian Trilogy – 3 Unforgettable Stories

Story 1.
"Lady Zero" High Speed Death—the Carrier of the Outbreak

Lady Zero is the Hypothetical Story which was shown by National Geographic television in November of 2005.

This November the deadliest viruses on earth aired on the TV special and it sited the mutation theory. "Lady Zero" is an unsuspecting female who is coughed on in the meat market and becomes actively infected with a lethal virus, likely Swine flu or Avian Flu Hybrid. She boards a plane and within 12 hours has begun infecting 12 others who will depart for different destinations. This one Chinese lady and the 12 on the plane that she infected will likewise, due to the DPT factors, cause **the acute infection of a *million* people in 2 weeks time!** (It will spread by means of other planes, hotels, subways, seminars that they visit. All this will happen before a vaccine for humans can be finished for public use!

The virus she carries could be H5N1, the killer strain perhaps of H5N1 and no vaccine can save her, and by the time a vaccine is created in 6 months, several hundred million people will die, her included. By the time she has traveled from China to London—within 14-16 hrs time, the infection of herself and the 12 that she infected on the plane will reach 1 million others within two weeks before a vaccine can even be started who will also die.

That being said; if one lady infects 1 million people in 2 weeks, then 1million may infect up to 1 billion people in the next few weeks or months. **This could quite potentially end the world.**

Story 2.
SARS China's Catastrophe

Example 2. True Story
Recently in the late 1990's the simple *flush of a toilet* on an upper hotel floor in a hi-rise China hotel infected 122 people. The fecal matter was infected with the SARS virus. Hospitals were overwhelmed with other sick people from all over the city. If this happened in the Sears Tower 72% would die in a few days! 50%—72% are the minimal mortality rates accepted by health officials.

The SARS virus may possibly have passed though the plumbing pipes or seals.

Example 3.
SOUTHEAST ASIA
Cooked Chicken and a documented–Human to
Human Transmission Family Case Study.

Example 3. True Story.
This is the true story of multiple human transmissions, within a single family and to a hospital employee in Vietnam.
From CNN October—November 2005.

In October 2005 TV aired a documentary on Avian Flu
localized in Asia, and dealt with a "bird to man" and "man to man" transmission, this is an example of the stage 4 CDC pandemic and is the beginning of a minor cluster formation.

I believe there are more "mini-clusters" occurring, in Asia and china, because there is poor medical care in Asia, and a high element of economic secrecy in China, leading to the fact of dilution of data and the downplaying death rates of men and animals and even a "cluster" spread.

In Southeastern Asia, stages 3 and 4 happened.
An Asian family cooked an infected chicken and two of the three family members were near death in the hospital many months.

The actual story of three near death episodes include *a man and his son, his daughter, and the nurse.*

A man and his son and daughter were eating a chicken

And the son became acutely ill and was **in ICU-intensive care for respiratory support for 8 months, then sent to rehabilitation areas.**

The daughter:
Was hospitalized less than 3 months and went home long before her brother ever recovered.

The nurse:
taking care of the brother, unfortunately became acutely – dangerously infected from being coughed on by the boy who ate the infected-cooked chicken. She too was hospitalized. The boy went home after almost a year of care.

The TV broadcast **showed the family again, eating chicken-or duck as soon as they got home! An interview with a local indicated that she would not change her eating habits for anything.**

People take their pets to talk with the wild birds in the town squares in China and Asia which leads to DNA mutations.

But, strangely **the father, who was apparently positive for H5N1** *is still asymptomatic.* Most likely he had been developing a chronic immunity to the H5N1 from eating infected chicken for many years.

I am worried, and so were the medical people on TV, that If the man was on a plane with others, and he coughed on the plane, the virus would recirculate through the inadequate filters creating a "Lady Zero" infection sequence, with an apocalyptic disaster! The use of our new Betadine™ proposed system would avert airplane infections.

So, theoretically if the asymptomatic carrier coughed on you, you could easily become infected with H5N1 and possibly die from it,

but if a Betadine™ was used cleverly no one would die on a bad plane or train or tram or hotel or hospital ,astrodome, or doctor's office. After the "man" breathes on the plane or coughs on you, within few days (in the absence of rare diagnosis and ICU intensive care treatment), you will die quickly, up to 72% of the time.

How this Vietnam Infected Family story ended and how it can be prevented in the future. The father may never have been antibody screened! or at least never revealed. The father of the family was never sick so it's possible that he can still be blood tested and save the scientists some immunity dilemmas, and deaths. No one had the capacity or sense to test him, or forgot too test him.

What is sad and common in poor countries that barely have hospitals, is that they have to burn the dead before an autopsy or tests can be performed.

No blood tests!
It is a misfortune that the asymptomatic father cannot be found now since he is immune, and he and the woman in Thailand are the only active human carriers, currently available to the press to locate, and we may have forgotten to test them inside and out! Many active and retired immunologists would love a crack at the Vietnam man's blood antibodies, immunoglobulin and dietary background.

Asian Man and Lady Zero.

It would not be good for mankind to allow the man from Vietnam or the lady survivor from Thailand to travel on a plane until they can be blood tested since they may possibly be recreating the lady zero story in real sense.

Critical imminent danger exists in Asia where DNA is traded like a street fair from wild birds to domestic birds.Recently the Chinese are using inferior vaccine to inoculate the chickens—which predisposes the birds to make horrible viral mutations. There is a small but real probability of a lady-zero coming from the man in Vietnam or from anyone in China and an entire family transmission—becoming larger making the smaller "cluster" cases into larger cities, states and countries. There is almost no way to study the virus where it began, and the Avian flu is killing millions of birds in some countries like China, Vietnam, Southeast-Asia and Turkey. Like a fire spreading, these countries have widespread poultry disease. Now Poland has "chicken-police" and France and Germany have had outbreaks in poultry in February of 2006

China killed 43 million chickens toward the end of 2006-07 period to stop the spread.

Absent Medical Data In Asia

Data is missing because the dead are burned quickly to prevent spread. China has been known to disguise its fatalitiesAll known viruses in last 100 yrs are from China. Lack of accurate data collection No travel restrictions of the multi-thousand monthly flight passengers from Asian areas, China, Japan and Thailand.The final outcome of an unchecked human carrier can be the end of civilization.

Using Vietnams Real Occurrence as a hypothetical model for the Annihilation of the Human Race.

Consider the nurse infected by the boy in Asia. The doctor treating the boy writes with a pen, and if the infection touches the pen, and the pen is passed to the nurse,the pen becomes a carrier of the germ. If the doctor happens to rub his eye for some reason, he becomes infected and can die. Now, if he had also passed the pen to a nurse aide, who then boarded a plane for New York city, both the doctor who and the nurse aid who borrowed the pen are now "lethal lady-zero" carriers of H5N1 or something worse! And both are capable of inducing a pandemic

Geographical-Tracking of Avian Flu
How fast is H5N1 traveling, spreading?

In miles/month the Avian Flu is moving at 2000 miles per month based on bird –only flights as the human cases died or stayed at home. It is a high speed virus rate of spread of H5N1 December 2005 to March 25, [th] 2006. World Trip-Map at a Glance. In cases it seems to be about 1500 miles per month for a speed. The H5N1 traveled from Southeast Asia to Russia in 4 months. April to July 2005, a few thousand miles. H5N1 traveled from Siberia to Romania in 3 months, and this is cumulative total of about 10,000 miles in 7 months time. It made its way into Poland in January 2006, and to France and Germany by February 2006, another couple thousand miles. Iran, Iraq Turkey reported outbreaks in February 2006. Nigeria killed 43,600 chickens in February 2006. It reached England in October another 1000 miles. It reached Canada from Siberia probably in November. It reportedly reached San Francisco, November, 2005. The media hushed up a story of a small boy who was possibly infected by H5N1—lovebirds. The birds may have come likely from China or Southeast Asia.

It's traveling about 1500-2000 miles/month and has nowhere left to go except for South America and Australia, and the USA. It was noted as being found in Alaska, but unconfirmed this April of 2006.A most interesting question to me is **why and how it is not in South America? Nor Australia.** Birds are the primary carriers and may fly to a new sector or travel on a barge or freighter to get to a new destination.

TOOLS for the reader:

To assist in reading the outline" format" is a list of Public Health methods of disease control. If you learn these words now, then reading the newspaper avian news will be easier in the months to come.

How to Contain an epidemic, disaster.

A) Direct killing of the host
B) Direct killing of the virus
C) Diet
D) Masking
E) Medication
F) Quarantine
G) Work restrictions
H) Travel restrictions worldwide

SECTION II

MEET THE PRESS

November, 2005 to May 2006 , a 7 month update.

Since this documentary seven months ago, not much has changed to improve man's chances for survival, but to put this is a historical perspective read and compare the data from November 2005 to May 2006.

This section continues with Meet the Press documentary hosting the four world health leaders. Later in the book is detailed discussion of medical and nutritional remedies for your use which are new innovative and maybe the only chance we have.

Data published from Asian sources went unmentioned at this meeting.

Meet the Press reactions
November 2005

Summary of current status meet the press
November 20th, 2005.
Members Present on the TV show were:

Director of CDC, center for disease control
Director of WHOM, Dr. Michael Ryan

Secretary of Health, USA
NIH, Director of the National Institute of Health.

All members agreed that there would be no vaccine in USA for 3-5 years able to be ready for a pandemic and there would be a 6 month lag time for start up production, once the system is in place. **Others recently claim that 1 billion doses of Tamiflu are available with only a one year shelf life from May1,2006 until about April or May 2007** and that's it.

Some very basic non-vaccine recommendations were made; to avoid some public events and proposed community education. It was not until March of 2006, 5 months after this first show that the USA was advised to save food and water for **a 3 month** period. The recommendations which I concluded and others concur, is **about 6 months** of food and water are needed per person.

In 2006 the the amount of vaccine available in USA is 4.3 million doses for our 300 million people. Things have changed and more is available but now the Avian is mutated and mating with a Swine-Pig flu virus; and consequently vaccine for a bird is not the same as for a bird-pig hybrid.

That 25 % of population of France and a few other countries have vaccine on order – but it's promised and ordered and not ready to be delivered. Health care workers of USA will be the first to receive the 4.3 million doses which were available as of November 25th, 2005 Finally, no mention was ever made of the Hungarians developing a potent and effective using 40 year old method, making 2 million doses a month and it was licensed by China in November and ordered by Jakarta in Asia in recent past months.

There was one reported case of a child in Thailand getting Avian Flu from its mother who was nursing the child but the panel omitted the story of the Vietnam family infection . . .

Advice was given to enjoy thanksgiving turkey was given to the moderator and audience.

Strange booklets for sale on the internet falsely state that a person cannot catch Avian Flu from cooking the infected chicken, but this unchecked advice is deadly.

The Television press story had some incongruities with Vietnam cases by suggesting Turkey was fine to eat if cooked, while having no knowledge of the 3 near fatal cases in Asia from eating a cooked chicken was a bit misleading. and *cooking or boiling the chicken does not stop the infection.* I would like very much to have the CDC test-Microwave a H5N1 Infected chicken **to see if** the **microwaves destroy the virus.**

Avian Flu: Nearly Impossible to Kill.

The survival range is almost extraterrestrial and It survives from – 70*F to well above 450*F. This is the basic chemistry thermal range of the Avian Flu virus.

A Review of Pertinent Events since Meet the Press
Of November 2005.
From November 2005 to May 6th 2005 the following occurred:

1) The 2004 on-hand 4.3 million doses of Tamiflu is now being augmented by a new genetic clone with a maximum capacity of 450,000,000 doses over some months beginning in March of 2006.
2) New ways to make vaccine besides eggs is underway because there will be no eggs to make vaccine for even Avian flu except for the first 450 millon in number 1).

My Proposal for—a non-vaccine protocol, trans global protections and an alternative to vaccine use.

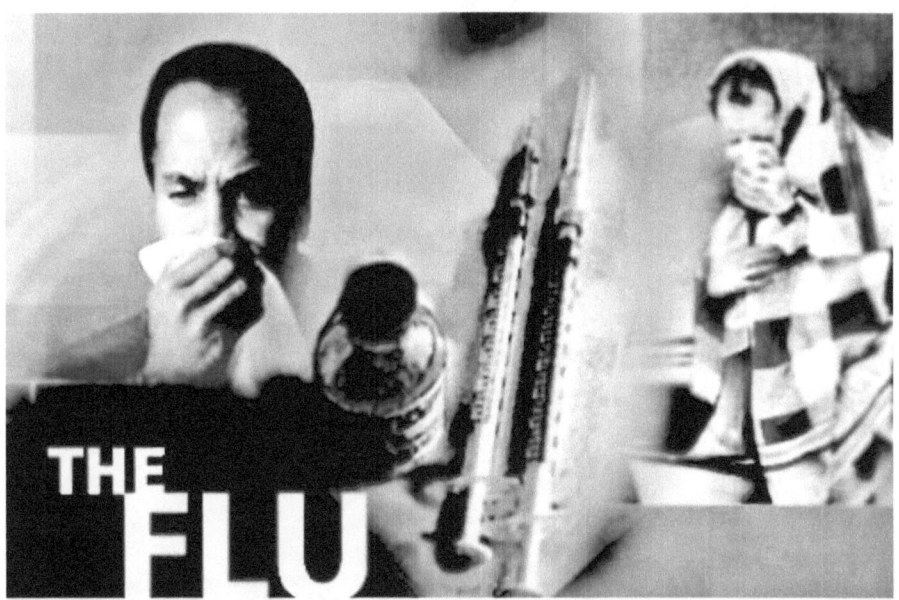

There is only enough vaccine for only 1/12th of the Earth's Population maximum output capacity announced in March,2006.

Winter 2005/2006:
Evan Vaccine / ap
President Bush outlined a $7.1 billion strategy in The midwinter months of 2005 and 2006 to prepare for the danger of a pandemic influenza outbreak, saying he wanted to stockpile enough vaccine to protect 20 million Americans against the current strain of bird flu. *20/300 IS 1/15TH of 100% needed*

How to Use the Advice Given in this book carefully
Is under the supervision of a medical doctor.

The following is an outline of my suggestions to public health people and needs implementation and testing. Aerosol applications to homes and buildings with Betadine™ or HEENT upper respiratory uses are still theoretical and are only with a doctors approval and only if a pandemic cannot be averted. Discoloration of the home and office will occur with use of the aerosol. But discoloration of office and home is not important when people are ill or dying all around you. *Discloration neeed not be eventful if the air is treated differently.*

Verified New Hope

The theory that I began in November of 2005 is now being supported by data from Japan from April of 2006. See the attached reference at the end of the book One saving factor is that the use of Betadine™ In HEENT orifices is highly safe and was and is used safely by knowledgeable doctors worldwide. **It is not just an "antiseptic";** it is a versatile viral-bacterial killer for many orifices of the body.

SECTION III

BETADINE™ IS POVIDONE IODINE

The sky is the limit . . . .

THE POSSIBLE ULTIMATE CONCEPT

CLOUD SEEING with ANTIVIRAL AGENTS.

The traditional method of cloud seeding for over 20 years has been the crystal of Potassium Iodide-KI. Specialized local dusting with betadine™ or PVPI in Safe sectors not over fisheries or fresh water can Stop the utter filth of the water in Katrina which contained Dead flesh, fecal mater, urine, cholera, ecoli and many More deadl metals mercury and lithium from batteries.

We are currently playing cards with a lethal virus with only one card in our hand – vaccine and this is not rational, since the card is not even an Ace! And the ace cannot be played because the vaccine currently being manufactured is only for chicken to man transmission.

Fish are killed by only 2ppm of Iodine (2 parts per million parts of water). But the fish are not eatable infected with VHSv viral hemorrahgic septicemia nor with Mercury levels so high as to Cause birth defects so in July of 2009 labels from the EPA Were apparently placed on all fish from Michigan with a Pregacy Warning DO NOT EAT MICHIGAN FISH—Hg-Mercury.

Research supports the theory that Betadine™, PI, can outlast any known virus up to 1997 and is famous for killing HIV in the operating room instruments and tables. It was sprayed in operating rooms from Japan to the USA harmlessly. In fact the Avian problem in Japan prompted the April release of data on H5N1.

It is not rational to approach airborne viruses with vaccine alone. Remove the stigma from Betadine™ There is a stubborn ignorant streak among the medical profession who when they hear the word "Betadine™ immediately categorize it as Rustoleum™— a coating, and they have no knowledge of its HEENT, neurosurgical oral and other extended uses.

It is sprayed in operating rooms world wide to kill everything in sight and used topically in the eye itself so it seems logical To allow more aggressive uses of PVPI (Betadine-to stop infection viral and bacterial.

Use the advice given in this book carefully And under supervision by a medical doctor. This is a **basic outline of public health and personal measures** and things the government should do for you and or you can easily do for yourself and family. *But the use of the medication in this book is ONLY to be used only in case of a pandemic outbreak not as experiment.*

The first studies of Betadine against Avian Flu was published as of April 22,2006.

Section III contains:

- LAVBAC-FM, my non-vaccine alternative to the uncertainty of vaccines
- Medical documentation of my theory
- Betadine™, chemistry and history nutritional aids and vitamins and agricultural programs for feeding animals experimentally
- Stages of a Pandemic (CDC)
-
- DPT Factors of spread
- Crop dusting of cities, and climate control with cloud

We have what we feel is a rational solution—

BETADINE™ POVIDONE IODINE.

A NEW FORMULA PROPOSAL
For testing hopefully at the CDC.

LAVBAC-FM is an abbreviation for the following:

L= Licorice
A= Anise, other food supplements
V= Vitamins
B= Betadine™ also called Povidone Iodine
A= Aerosol
C= Cloud Seeding & Climate Control
F= Filters for masks
M= Masks made from Betadine™

LAVBAC-FM will offer a new way to look at viruses.

The entire contents of the book is based upon a concept called, LAVBAC. This is an abbreviation for the following: Licorice, Anise, Vitamin Therapy, Betadine™ Medication, Aerosol Crop-dusting and Climate Control and of course Masks and gloves of the right quality.

An implementation of Diet change for humans and agricultural animals along with a clever and careful use of the medical chemical Betadine™ – Povidone Iodine in air and water will actually provide a strong level of protection against all airborne viruses (Betadine™ is now 50 years old & used millions of times daily world wide)

The letters DPT
refer to the vaccinations which were used for years for the 3 routine immunizations, namely diphteria, pertussus and tetanus. And the same medical letters "DPT" can also be used here to conveniently represent:

Density of Population
Proximity Factor,
Transportation Mode.

The 3-DPT factors drastically increase the mortality rate of Avian flu and other similar infections, reported to be as high as 50-72% in Asia. Avian flu in the past 10 years, traditionally has passed from bird to bird, and occasionally bird to man, and in 4 documented cases a human to human transmission occurred which is alarming.

If the human to human virus takes hold and the vaccine to protect us **will take 6 months to design and make effective, during which time a billion people** will likely die.

The Avian Flu virus is spread through the air in small droplet formations In some cases a sneeze or cough can carry many virus particles 15—20 feet depending on conditions and location. Its spread is the same as Hemophilus Influenza, Adenovirus, SARS and the routine common cold. One thing which makes it so devious is its size,—the size is 1u that is one micron, making it able to penetrate any mask on earth except for the 3M tested 0.1 micron mask. Most of the other masks do not block its small size. It is a wizard of a mutator.

The book discusses REAL ways you can protect a home or office with a chemical-medication, pending CDC confirmation or with your doctors advice. **Prevention and Containment is the BEST treatment.** Iodine allergy is a contraindication to the use of Betadine™ also called Povidone Iodine, but incidence of reactions to Betadine™ itself is less than 15 cases in 10 years. It is likely used individually well over 25 million times daily by nurses paramedics, etc, but the way in which it can be used and should be used has been long forgotten.

The long gap in loss of understanding of Betadine™ occurred starting about 1980 and the **ignorance** of its former uses in medical circles is prevalent and **dangerous** at a time like this,.

Even the average medical person has no knowledge that Betadine™ can treat adenovirus infections of the eye, sinusitis, and that it was sold as a mouth wash and even a Vaginal douche without complications. Additionally, it is used for Herpes Vaginalis, and HIV decontamination of surgical counters and instruments and operating rooms. Herpes Vaginalis has other medications, but what do you offer to a woman in the outback or living in jungle countries who has never seen a pill,— you offer a simple, cheap method like Betadine™

Betadine™ can be safely used on parts of the respiratory tract and that coupled with a good traditional "antiviral diet", may save your life. It is also used in the treatment of sinusitis by some ENT doctors, ear, nose and throat doctors. Ventilator support may benefit from using Betadine-PVPI

The *incidence of allergy* is minimal and one can put a few drops on the volar wrist and if no rash appears Then the person is not allergic to iodine. It takes about 20-30 seconds, so an entire stadium or city could be done in a few hours or a few days if it's the city! The theory is a simple extrapolation of the old days.—from the 1930's when Iodine was put in water of suspect origin.

Betadine™ and diet change might be all that is needed to turn the edge of a nasty, nasty set of airborne viruses including H5N1, Ebola, SARS, and even Smallpox. *It's the **"drop in the bucket"** of your vaporizer which will save your life more than likely if no vaccine is available.*

Certain methods discussed here are theoretical and need permission from your doctor or the CDC to implement, but enjoy a real way to save you and your family's lives.

A Bird's 'Eye-View' of the problem

Fact: In 2006 this was the comment:

There will be no vaccine available to US general public for many months – it has been predestinated for use in the health care workers, police and firemen and government officials.

The available vaccine is **only for the chicken to man type virus, not for the man to man virus.** Production is slow and preordered and backordered. Even Tamiflu is backordered 3 months in local hospitals, and it may be 3—5 years before a useable vaccine is available just to meet the needs of the USA . (Meet the Press November of 2005).

There will a shortage of about 3 million hospital beds if only 25% of the USA is infected. Gymnasiums will be used for medical care! The hospital respirator shortage is always estimated a shortage of 25,000. But, I consider 25,000 to be 1/10th of a needed 250,000 ventilator beds at minimum if functioning is possible.

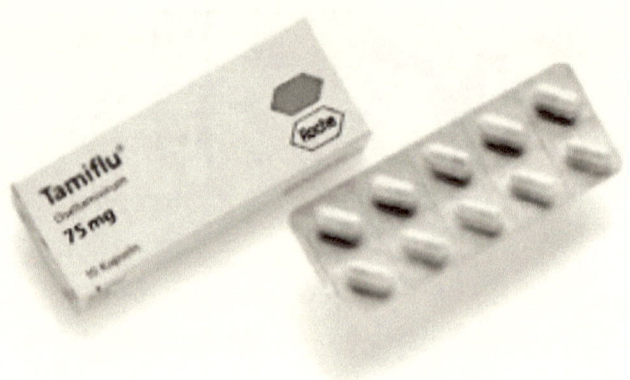

Tamiflu™

Costs about $ 75.00 /week and is short on supply and not effective in human to human transmissions. It is very effective in bird to man infections and saves lives.

There is another way to augment Tamiflu's coverage.

This is a factual guide delving into a forgotten medication and offers an immediate solution to the problem at hand regarding H5N1 and the other killer airborne viruses, such as Ebola, Small Pox , SARS,etc.

The plan suggested in this book helps minimize the threat of not having a vaccine in the event of a pandemic outbreak, such as using Povidone Iodine as an upper respiratory medication and vaporization. Betadine™ is also called Povidone Iodine.

The edge can be taken off airborne viruses with Betadine™* in ways that we forgot existed and in **new ways** Enclosed is a scientific protocol for prevention of Avian Flu, etc.,which is affordable to everyone on earth, regardless of profession, and is 1/1000th the cost of Tamiflu and similar medications out there today. It may offer some **modified form** of medical protection which up to now is not mentioned. The principles taught here are applicable to all airborne viruses and some chemical warfare agents, not just H5N1.

Our Theory is plain and simple.

Go with what works! Betadine™ is a Standard of Practice medication— SOP, and make use of some of its special applications which are not well known, but may be applicable to the current Avian problem. The Plan covers past-present and future calamities of virology. For future reference, you will see the term HEENT used quite extensively in this book. **HEENT** refers to **(Head, Eyes, Ears, Nose, and Throat).** Remember doctors use it on preoperative areas, on the brain itself in the eyeball, in the nose, intravaginally,and Intraorally as a mouthwash.

Betadine™ has been used medically on all HEENT systems in Head and Neck Surgery for 50 years, will likely kill H5N1.and maybe Ebola. It has been used safely in every orifice of the human body.

Points of consideration

Nutritional remedies:
3000 year old substances are available for men & animals alike. *Licorice, Anise, Pine Needle Tea,and Birch Bark are ancient methods of dealing with illnesses, dating back to the Egyptians and Greeks.*

Medical Remedies:
Vaccines must be used if available and proven.
Antibacterials such asBetadine™-Povidone-Iodine is half a century old. Situations for the use of Povidone Iodine are:

- Environmental concerns, killing a virus before it infects
- Aerosol for homes buildings and small cities.
- "Crop, city-dusting" of clouds over cities, counties, states, countries and continents.
- Implementing improvements in air and water transport systems with Betadine™

Consideration needs to be given as to the treatment of humans and animals with a non-vaccine medication, diet, and a revision of all heating and cooling systems worldwide, by implementing Betadine™ Other known nutritional elements form an encompassing plan which we can offer some *real and instantly deliverable protection to the earth in all areas* of concern. Something like Betadine™, and LAVBAC nutrients may fill in some spots in the void of no vaccine for 11/12ths of the earth's population.

Diet alone will not kill Avian Flu, but I wish someone would take a stand on recommending a few standard fundamentals, like Vitamin C, Zinc, and Selenium, and even some ancient herbal teas if their use is tested as safe.

Using Betadine™ with Licorice Anise Vitamins and Climate control creates the word LAVBAC. LAVBAC may stimulate new way to think about protection from massive viral and bacterial infections as occurred in the New Orleans Hurricane Katrina. Feed LAVBAC components to animals and make them less vulnerable to viral attacks.

The principles discussed are applicable to other and future calamities or diseases, and possibly the respiratory aspect of the Povidone medication may even be helpful in chemical warfare.

*Feel free to criticize or correct my thoughts, but at least substitute your own and entertain or share the notion that we really need to think and push and spend a lot of money on some serious **Non-Vaccine methods** to save our lives.* So if it is not Betadine™ , then perhaps it will be a synthetic chemical-medication, and if it is not licorice, then something else, but lets really try harder to feed test animals and humans something to stop airborne viral killers before they stop us.

BETADINE

The forgotten medicine
This information is to help the reader understand Iodine of the Old Days, and Betadine™ or Povidone-Iodine of today.

A clever company Purdue Fredrick Invented Betadine about 50 years ago, by hooking a painful Iodine molecule (medically helpful but painful) to a gelatin like material called Povidone, or polyvinylpyrrolidone iodine PVP is the polyvinyl Pyrollidone and I is the I2<-> 2I—complex.

Povidone is basically harmless, and it passes through the body unnoticed kind of like a soothing gelatinous-chemical Carrier keeping the Iodine from burning skin or mouth linings It is not meant to be ingested but it would take much much More than 4-8 ounces to cause a critical condition in a full-Grown man.

The few drops put on the wrist of a prospective user Is positive if itching and a rash occurs and is very very Rare,,the PVPI is used billions of times daily and rarely More than a dozen adverse effects to eyedrops alone from 1995-2005 only 15 reactions occurred to worldwide use of eyedrops in 10 years! PVPI is great for viruses of eyeballs themselves where it is often used against airborne viral infections.. ex. adenovirus

This is a Molecule of Povidone Iodine and it was and is used In airborne virus infections of the human eye, and in brain surgery.

This is a drawing of a molecule of Povidone—Iodone
[—CHCH2—]—
   |
  N
   ^ O
 C C// *x *I* (Iodine)
 C—Cn

POVIDONE-IODINE is BETADINE™

When doing brain surgery during the Golden era at the University of Michigan, the pioneer in human Neurosurgical Techniques – Eddie Khan used Betadine™. He actually poured it onto the raw brain surface intraoperatively. This protected the patient from any and all infections. This is a statement of fact and observation. Betadine™ has been used for half a century and has been used internally and externally safely in surgery.

Stages of a flu pandemic

The world health organization (WHO) has developed a global influenza preparedness plan, which defines the stages of a pandemic.

The phases are:

Interpandemic period[1]

Phase 1: no new influenza virus subtypes have been detected in humans. An influenza virus subtype that has caused human infection may be present in animals. If present in animals, the risk of human infection or disease is considered to be low.

Phase 2: no new influenza virus subtypes have been detected in humans. However, a circulating animal influenza virus subtype poses a substantial risk of human disease.

Phase 3: Pandemic alert period is human infection(s) with a new subtype, but no human-to-human spread, or at most rare instances of spread to a close contact.[3]

Phase 4: small cluster(s) with limited human-to-human transmission but spread is highly localized, suggesting that the virus is not well adapted to humans.[4]

Phase 5: larger cluster(s) but human-to-human spread still localized, suggesting that the virus is becoming increasinglybetter adapted to humans, but may not yet be fully transmissible (substantial pandemic risk).

Pandemic period

Phase 6: pandemic: increased and sustained transmission in general population.

1. Notes: the distinction between phase 1 and phase 2 is based on the risk of human infection or disease resulting from circulating strains in animals. The distinction is based on various factors and their relative importance according to current scientific knowledge. Factors may include pathogenicity in animals and humans, occurrence in domesticated animals and livestock or only in wildlife, whether the virus is enzootic or epizootic, geographically localized or widespread, and/or other scientific parameters.
2. The distinction between *phase 3, phase 4* and *phase*
3. *5* is based on an assessment of the risk of a pandemic. Various factors and their relative importance according to current scientific knowledge may be considered. Factors may include rate of transmission, geographical location and spread, severity of illness, presence of genes from human
4. strains (if derived from an animal strain), and/or other scientific parameters.
5. This is the current official WHO alert level.
6. There are many who believe that the current situation in Indonesia warrants this level.
7. Source for this list is centers for disease control and
8. prevention of world health organization and wall street journal October 22,2005.

SECTION IV

SCIENCE—LAYTERMS

Your Worst Nightmare and Enemy—the Virus.

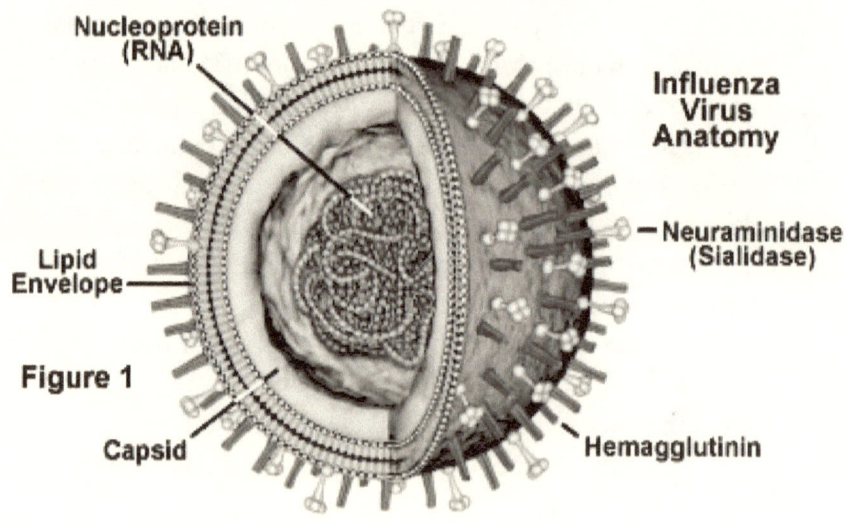

Nucleoprotein (RNA)

Influenza Virus Anatomy

Neuraminidase (Sialidase)

Lipid Envelope

Figure 1

Capsid

Hemagglutinin

How big is avian flu?

*If the picture of the virus above was the diameter of a human hair, then the size of the virus would be this big = * , the size of the asterisk mark compared to the above drawing.*

This virus in real life is 1/40th the size of a hair on your head, or 1/1000th the size of the tip of a ball point pen!

It is 1/one-millionth the size of a yardstick or meter.
Being this small it penetrates over 90% of all world masks.

Comparison of Bacteria to Virus: A Simple Explanation

Bacteria have cell walls, rather simplistic compared to viruses which are complex. Viruses are geometric and have only coatings or envelopes or those with coatings and those without a protective coating. The envelope is made of a lipid layer. (Lipid means a fat-based chemical.) Avian flu virus is an "enveloped" virus. Betadine is more effective against non-enveloped viruses but does well against –enveloped Avian Flu.

HOW BIG IS YOUR ENEMY?

Avian Flu virus is one one—millionth of a meter or a yard stick! The virus size is about 1u, one micron. A micron is 1 1/millionth of a meter. This means H5N1, **and most viruses are one/ one thousandth the size of a pen point or 1/40th the diameter of a hair on your head! The enemy is fast and small, penetrates most** masks world-wide! The masks you see Asian countries being worn shown on TV news have no effect against the average virus.

Tuberculosis is on the rise again and is perhaps the strongest bacterial customer on the block.

Coincidentally, it was discovered in 1997 that Betadine™ was effective against the non—enveloped and but still kills Avian Flu H5N1 non-enveloped but it is still the best drug against the rest in the study. *Viruses do not respond to antibiotics!* The only known antibiotic which seemed to act against "flu" years ago was erythromycin, subsequently now called Zithromax or "Z-pak" but, I still wonder if a variant of this is useable for the current problem, but it is no longer used for influenza.

Cross section of a virus

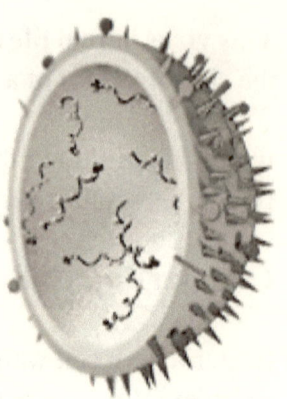

Avian Flu virus is an enveloped virus, which can mutate at any point in time. It has transferred from animal to man, and man to man consequently putting it at a semi-critical stage 4 of 6 stages leading to potential oblivion of 1/3 or more of the human race in less than a year's time. (National Geographic television special November, 2005.

The best explanation of what a virus is in included in the references, namely the Indian Journal of microbiology, S Phandi.

(See references for how to locate the article.) This **Indian article** is the best detailed and easily understood concept of viruses that I have ever read. (2004) See references.

What is a mutation?

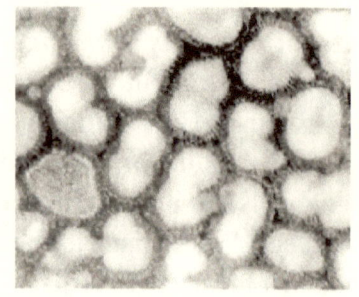

A mutation is a change of identity, a disguise as it were like changing routine styles in the middle of a dance, first a waltz then a tango a different identity. A virus can change its chemistry at will, but bacteria are more " normal", reliable and treatable infections. In layman's terms the simplest explanation of mutations and vaccines is to think of a virus as being a car, and the vaccine as being the insurance coverage.

Vaccine and car insurance analogy.

This is a simple explanation of why vaccines are risky and unreliable:
A stable vaccine cannot be made until the killer mutation is actively killing large numbers of people and the set-up time to make a new "killer virus vaccine is 3—6 months, by which time a billion people could die

Since the Avian flu scare Of 2005 and 2006 the avian flu has enabled its genes To pick up attributes of pig-swine flu,animals and human Receptor attack sites used by the Swine flu. So, instead Of being bird to bird, and rarely bird to man. The virus Either one can go virus directly to man. V→MAN. Smokers know how to manipulate to be near a smoker To borrow so allow the virus to be as clever as a smoker.

You cannot buy "Vaccine"-Auto Insurance unless the auto exists to insure, unless you know what car you're buying. The killer strain has not evolved yet, some think it will within 3 years, others do not believe it will occur at all, but the **immunologists are very worried and do believe it will have to happen in the next 1-5 years.**

If you drive a ford car, and you are covered by an insurance policy. You are insured for that model of car, not a truck or a van or a future planned vehicle. A vaccine made for Swine flu Cases is for Pig to human transmission amd just wont work, and a man to man swine vaccine is not going to work either , until the swine virus kills in clusters 100-1000 and then its' 6more months to recreate a virus which can kill this customer: a swine to man, man to man cluster former infector of 1000 people at **a time vaccine.**

The virus is lethal by changing its look and your insurance (a good vaccine) does not cover the 8 cylinder high-speed virus! To a vaccine corporation it must be like insuring drivers trading cars at 60 mph.

It is *impossible* to create a protective "insurance-vaccine" for a killer human to human virus which does not exist yet. The proper " insurance-vaccine' cannot be **made** *until the killer virus exits or surfaces gets a name or identity as the killer and is beginning to kill many 100-1000 clusters at a time*

The killer mutant strain by definition will have to begin by killing 100-1000 people at a time in order to be defined as the killer strain it must happen in multiple geographic locations.

Current Vaccine Insurance Coverage:

Current H5N1 coverage is a vaccine is for chicken to man primarily and the vaccine is insurance for the current H5N1-chicken to man-(model car). The vaccine has no effect on the man to man-virus (truck-van virus) which has not evolved yet. *If H5N1 changes to a truck/van model, then the current vaccines available to us are 100% USELESS!*

Viruses can mutate and change their geometry, especially when chickens are vaccinated with cheap vaccine which happened in China November 2005 **predisposing to massive mutations!**

A vaccine is specific, like car insurance, **you cannot trade and switch insurance policies during the accident.** If H5N1 is now a **car**, and it can instantly become a **truck** so we are currently **all "driving naked"** uninsured! *The current vaccine is incorrect for what will possibly happen.*

The wrong or absent vaccine will cause upwards of 1—2 billion people to die before an effective vaccine can be made; that is 6 months will pass where the world has no adequate coverage-that is—insurance. There is no choice in this dilemma; we are naked of any reasonable protection.

Fact:
Current vaccines manufactured by Roche, Hungary and China may have **not been tested on human to human transmissions sufficiently to make a statement of a cure rate.** The human to human cases that do exist are about 9 World wide and they have absent medical records. Tamiflu™ has been used successfully on those who ate infected chicken, but I have seen **no** documentation of human to human cases and Tamiflu™.

Time frame for the doing biogenetics of safe vaccine
It will take a minimum of "6" months after a Killer Strain evolves for a useable vaccine to be produced. This was announced on a CNN television show in March of 2006.

Another possible disaster is having our vaccine come from infected eggs.

Two High-Powered Examples of Biogenetic Research Development of the new Ebola Vaccine and new H5N1.

For example:

Design of a vaccine against Ebola was announced in November,2005. This is the only recent success of making a killer safe vaccine. It was designed at an army research ICU base central USA. For Ebola, the army made the creation of a" monkey—safe vaccine" for monkeys only.

The monkey-safe vaccine method:

The Ebola vaccine was piggybacked "Trojan horsed"—snuck into the monkey's immune system by using a standard "flu virus" and attaching the Ebola virus to it, thereby sneaking it into the monkey.

It actually protects monkeys from dying from Ebola, which is the deadliest virus on earth, but is self-limiting because the people bleed to death so quickly within days so that the virus dies with them as long as the body is incinerated, and disposed of quickly.

Development of the New H5N1 Vaccine this past winter.

The current Breakthrough of a new vaccine to protect against H5N1 involved splicing a chromosome tip from the H5N1 virus and mating it with another common flu virus using a procedure similar to the Ebola army vaccine for monkeys, But the H5N1 strain that was "clipped" was not for protection against the **killer human to human virus strain** which has yet to mutate and is leaving some virologists extremely worried. The human to human killer strain does exist to be tested in Asia and perhaps is in secret vials in research centers.

And even if it –the new "mated-genetically clipped" vaccine did work it would not be able to be manufactured except for about 1/12th of the earth's population, leaving 11of 12 people UNVACCINATED in a Pandemic Phase leading to the death figure currently Comments of 3 billion fatalities worldwide are now accepted in intellectual circles of the virologists.

The events of 2006 with the Avian Flu are summarized here but looking at the news this June 2009, the scare of the Avian Flu did nothing to 1) supply an adequate number of masks Nationwide, 2) had done nothing to observe or enforce pig hygiene if there is such a thing, 3) nothing with the Swine flu was improved from a point of Preparedness, no masks and no vaccine, and a lot more dead people. So history did not Make us smarter. Refresh your memory of Avian mayhem In November 2005-2006. Remember this book recommended And still recommends the very same Betadine or PVPI as useful in agriculture, cities, homes cars,planes and boats.

The End Point
No chicken and No Egg.

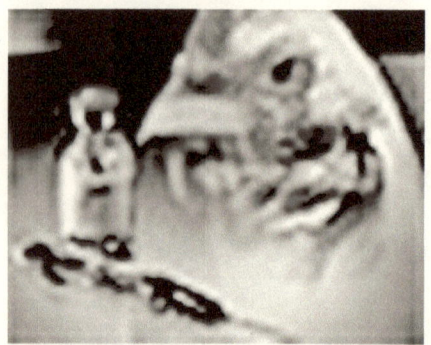

How to see danger—point blank!
It is the end of the line for vaccine production.

STATISTICS TO BE KNOWN.
The current plan of the entire World has all these
major weak spots:

We are sitting ducks for another virus. (2006)

    OUT OF EGGS
    OUT OF VACCINE
    OUT OF LUCK
    ECONOMIC BREAKDOWN in infected and normal
    If you cannot eat you cannot work nor fight a war!

1) Right now there are only 4.3 million doses of questionably effective *vaccine* at the disposal of the USA which is set aside for the health profession field primarily to be able to continue research by keeping the sick victims alive. The doses were increased after the events ended in 2006.

The newest cloned Avian vaccine can achieve a level

Of 450 million doses, but that is the finite limit; after that The man and the world will run out *of safe eggs to produce vaccine.*

2) **If we are out of eggs for Avain Flu vaccine production,** then by what method do we prepare vaccines for the next new virus.

3) *Work shortage prediction* by China was estimated at 35% in less than two months if pandemic occurred. So, we need to know what and how we are to work from home. If you cannot eat you cannot work nor fight a war!

4) *Loss of the poultry business will* put a huge segment of the population out of work—at ends, the producer and the consumer end.

5) If *mad-cow disease* comes back with Avian Flu we have arrived at shutdown economic levels.

6) There is **NO MEDICAL NON-VACCINE-PROTOCOL** for people in case of a killer mutation for fear of alarming the "public". The only suggestion thus far is to save food and water which was just released publicly in February of 2006,

7) There are *no blood studies of 4 human-near lethal cases* documented in a Southeastern Asian village regarding the human-human transmission CDC stage 4, H5N1 virus.

VACCINE AVAILABILITY AS OF MARCH 12,2006.

TYPE OF INFECTION	VACCINE AVAILABILITY
Chicken to chicken	China yes, and some from Roche
Chicken to man	USA 4.3 million doses., More became available just before Christmas 2006.

Chicken to Man Newest Vaccine.March,2006.
World announcement of "New gene-clipped" vaccine Maximum 450 million doses and then there will be no eggs left to produce any more vaccine for anything on earth.

Tamiflu™ **has about 1 billion doses with a one year shelf—life for use on suspect cases** and it is not designed for a human to human transmission mutation virus but is useful with the QDEL TEST for Influenza A and B treatment protocols.

TYPE OF INFECTION	VACCINE AVAILABILITY
Man to man	none none

(not made at all and will take 6 months to produce after the first clusters occur cases of 100-1000 cases must happen.

Killer Virus-Vaccine Gap existed in 2006 Avian flu, and again in 2009 Swine flu, and fortunately for all of us the swine flu did not mutate as it did in 1918, fortunately! But will anyone stop crowding the Pigs, or use PVPI or the Screening test for Avain flu DNA from California . . . //// . . . //// . . . //// . . . .

This is a scientific example of a mutation crisis:

A 6 month "gap" discussed on Meet the Press November of 2005. It explained that if a new virus "killer type evolves, it will take up to 6 months for a new vaccine!

So if a new virus like the "killer "1918 Spanish Influenza" occurs on this 4th of July 2006, mass casualties will occur "untreated" and continue for 6 months and progress right up to the single first dose of "Human Vaccine" comes at Christmas time in December of 2006, but the deaths will continue for about a year total and many people will die approaching the 3 billion plus mark. Even with inoculation using 450 million doses of the exact Proper Vaccine, it is only going to cover **8% of the earth's population at maximum production capacity or 1/12th if put in a fraction.**

These are facts from the man who cloned the new H5N1 vaccine on National TV in late March,2005

The saddest part is we have tunnel vision and cannot see that other China viruses are more frightening than this in the proper setting.

SARS scared the Chinese medical community; it overwhelmed them.

The DPT factors

Density of Population
City living and work setups vs. rural.

Proximity
Locations conducive to intermingling of people,socially,domestically and work settings such as cubicles, libraries etc.

Transportation
Travel modes in use and the country.

Some 'virus' re;ated transportation facts 1918-2005.

1918   walking, trains horses

2005   walking, trains, cars, buses, subways, planes, space.

Space Travel (What comes in from our space trips?)

The DPT factors can help predict mortality and morbidity rates, hospital needs and can be used to predict the speed of the disease. The mortality rate depends a lot on the type of travel used such as planes trains, cars, subway and buses, living and work conditions. The density, crowding mingling and travel modes are 30-50 times greater than in 1918 and are very important considerations.

Example:

A recent autopsy of 1918 plague victims proved to be flu H1N1 and it killed 100, million worldwide. It killed 30 million plus in the USA alone.

The population now is 3 –4 times that of 1918.

But it is **not a simple matter of multiplication**

Off a 100million in 1918 x 3 or x 4 for the population now. It is far more serious!

The density factor is probably a fair 25 times greater now

So that 100 million dead in 1918 could be calculated

As 100 million x 25 = 2500 million or 2.5 billion which is

The currently accepted death estimation for a major outbreak of about 3 billion deaths.

Anticipated deaths if an Outbreak occurs this year Is 2—3 billion people and that is $1/3^{rd}$ to1/2 of the earth's entire inhabitants, not counting the dead animals!

The mortality rate in Kansas farmland is going to be far less that in New York City.

DPT—Applied in Real Life Settings
Lethal set up in ventilation systems.

Depending on where you live and work can be big trouble, for and impact your home and office setting and the workplace. For example take the former World Trade Center of 911, where 50,000 people worked in two buildings sharing similar ventilation and plumbing systems: This arrangement of close proximity—density of office quarters of one person to another was a bacteriologic Petri dish waiting to be inoculated. Think of the virus as a lit match being dropped into a pool of gasoline. Such a system with airborne viruses around will kill 50-72% of 50,000 people in a heartbeat unless some other method exists to protect the home and work environment. Think of it as a lit match being dropped into a pool of gasoline.

If Avian Flu or **SARS** infected the Sears Tower, in Chicago the mortality rate would run up to 72%! Death in 72 hours. The current "DPT" factor in urban areas is "inviting" and is dangerous for a pandemic of any infection! The routes and the rates of transmission are infinitely greater and incalculable! The method of spread in 1918 is a dinosaur compared to a Lear jet in 2006

Economic Disaster, East West Trade Imbalance:

The greatest danger imaginable is a shutdown of the poultry business or pork industry which will happen like fire beginning with the first death of a chicken. China just killed about 43 million chickens in the winter months; and we may be doing the same or worse.

SECTION V

BETADINE™ DIVERSITY.

The Multispectrum uses of Betadine™, theory and practice.

This is only one unused facet of the medication.

Cloud Seeding with Antivirals and Crop Dusting

The use of PI-Povidone Iodine in crop dusting of major cities contaminated or destined for infection with deadly viruses or microbes is imperative. PI linked with water treatment and a clever use of medical-chemically treated masks will also save lives. Let's get busy and synthesize a powerhouse relative for Betadine™. The Sharps Plasma Ion Exchanger claims to kill Avian Flu, but is very expensive For citywide use can could be used in conjunction with a chemical concept will really help.(Sharps Plasma Ion Exchanger)

Using this medicine against the Avian Flu, etc. could be managed by a doctor and incorporated into all existing 0.1u masks and filters. This proposition must be studied and tested and proven **quickly** by major industry and all world governments.

New Uses to be tested and implemented quickly. General Principles of Use.

Traditionally, Povidone Iodine was a topical disinfectant, *but, Betadine™ can be safely used in or on: all orifices HEENT head ears eyes nose throat vaginal, hair, rooms, surfaces and air itself.*

The strong points of Betadine™.

➤ Betadine™ can be used on humans, and homes and buildings and countryside's is a doable—a global plan for prevention & reduction of pandemic spread.
➤ Iodine allergy (shellfish) is rare, and is contraindicated for the use of PI,Betadine™; however, one can do a skin test using a few drops on their wrist. If no rash appears then it is safe to use. The concentration in buildings and homes of iodized air may be effectively veridical and not reach **minimum lethal dose** levels for the inhabitants.
➤ This plan is as affordable for a Mongolian mountain man or a NYC worker.
➤ Iodine comes from mines and it is abundant.
➤ Betadine™ is cheaper than paint by the gallon!

- ➤ Medically proven — It has many applications and forms of use, human applications include rooms and buildings and large farm and barnyard areas.
- ➤ Nutritional : Using anise and licorice which have been used for 3000 plus years in Egypt Greece ,Turkey and Arabia, will be helpful, surprisingly economical and readily available and plentiful.
- ➤ Anise, licorice are sold here in the USA and are quite helpful in antiviral or medicinal properties for centuries.
- ➤ Vitamins and minerals need to be used,
- ➤ Current public health and safety measures may insist that strict travel restrictions will be imposed on all the USA. **about, sincerely considered and tested.**
- ➤ *It is simple: kill the virus environmentally before it multiplies.*

Goals and Suggested Health and Governmental Testing

Objectives of this section suggest that in the absence of any medical/ chemical alternative to a dwindling vaccine supply, it would seem logical and sensible to try this proven medication, or a new synthetic or natural variant of the same. The lack knowledge of Betadine™ or PI as **much more than a topical** disinfectant prostitutes its successful usein Adenovirus cases of the eye, in the treatment of vaginal Herpes, and nasal packing for sinus infections. Povidone iodine may be the only card we hold in our hand against the vaccine gap! So try to implement the sensible suggestions below.

Discussion of items in this section include:

Hosts
-Animal
-Human
Medical Helps
Nutritional Helps
Laws –permit to travel issued, restrictions
Medical aerosol prophylaxis

Beta Rain climate control
Cloud seeding with Betadine™

Direct Applications of the Above Plan.

Betadine™ can be used on humans, and homes and buildings and countryside's and its use is an economically, doable global plan for prevention and reduction of pandemic spread. Try to implement this very sensible, cost-effective plan.

A) **Treatment of animals** and immediate change of their **diet** using anise and licorice or similar additives experimentally with CDC testing and the department of agriculture.

B) **Animals** being treated with **Betadine™ aerosol** by means of aerosol in coops and barnyards and water sources. The American farmer has been using Betadine™ for decontamination of hands for many years. The American farmer is a smart cookie!

C) **Man** has been being treated with medically proven Betadine™ constituents, for the head-ears-eyes- nose and throat successfully for half a century!

D) **Nutritional aids**, foods vitamins and minerals for proposed use. Some suggestions are licorice, anise, and vitamin C.

E) **Housing protection**, work protection, such as duct tape for windows, water and food supplies. Teach anyone anywhere how to package up a simple "flu kit" for their own use and protection.

F) **Permit** to travel from house to work laws, with permits primarily issued to health care workers, and police and firemen. Travel except for vital personnel, will be restricted.

G) **Aerosol** use of Betadine™ for people too, for example on barnyards, and schoolrooms, and homes, should be strongly considered. **It is commercially available world-wide as an aerosol, and no prescription is necessary!**

H) **Cloud Seeding** with Betadine™ to cover major cities, etc. Salt trucks seeded with Betadine™ could possibly be used to cover major areas easily.

I) **Affordable alternatives** must be given to the people to be used with or **without a vaccine** and permitting a world protocol for men and animals. Do **genetic alterations** like those done in England that yielded a gene alteration making the given chicken immune to H5N1, but it's a long road ahead, but we do need a new chicken and a new egg!

J) **Shutdowns of the workplace.** Provide for and protect against massive "shutdowns" of major food And non-food industries. If no one is able to work who will dispense paychecks?

If driving is limited, who will shop and how?
If the birds mate with the American pets what happens?

China in October 2005, predicted a 35% work loss if the flu mutated; so how will they or anyone else function on a 2/3rds work force?

Global economy will fall apart within a few months of an outbreak, killer mutation.

> *Fact:*
> *People may not be allowed to leave their homes, much like the World War II days, except on specified days to shop for food . . . and only on days of odd or even street addresses! They will be required to wear a mask and gloves into the stores if a mutation stage 5 occurs. Oddly enough there are not enough gloves or masks to service America!*

Good references to see:

- Wall street Journal October 22, 2005.
- The Indian journal of microbiology Is referenced at the back of this book.

This section and A-J above will teach you a new way to think regarding surviving a viral outbreak, and urge you to ask health and science officials to think 'non-vaccine'.

Also try to remember there will be a **serious shortage of hospitals in USA** if the flu attains a 25% infection rate and the real number of beds needed is about 9 million we are short 3 million beds, and some of bed shortages will be remedied by using gymnasiums or shelters cots and bedrolls for patient care.

Medical terminology

It is important to understand the following terms.

H5N1:
H5N1 is Avian Flu's laboratory name. It is an airborne virus, like Influenza A and Influenza B, & Adenoviruses, Corona viruses, and SARS.

Upper respiratory **means head—ears—eyes—nose—throat entry to chest—HEENT is a medical abbreviation of these entries.**

HEENT
H head, hair
E eyes
E ears
N nose
T throat
T topical, preoperative choice and minor wound care.
V-vaginal (is an unusual use of Betadine™ for Herpes

Betadine™:

- Betadine™ was used almost exclusively for hospital counter tops to stop HIV. A few years back. and puncture wounds in operating rooms, even in the treatment of Herpes vaginitis, as a vaginal douche. It is also used in ENT surgery, sometimes for nasal packs, and comes as an eye drop for surgery of the eye, etc.

Airborne viruses are upper respiratory tract invaders

All of these sites in HEENT must be blocked to stop an Avian infection, or other airborne viruses.

(Ears are exempt unless a person had an ear infection.)

Viruses penetrate ordinary masks and clothing like a "22" caliber bullet—they are about 1 micron in size.

One micron is about 1/1 millionth of a meter (yard).
or 1/40th the diameter of a single hair on your head.

Hosts:
Host means in whom or where the germ lives, who or what harbors or holds the infection. Now in Asia and most of world, *the chicken and ducks* are the hosts or carriers, and also some men and women in Southeast Asia are reported, up to 4 human to human cases in stage 5 are documented, and the rest of the dead are burned so no medical records exist. There are many more than 5 human to human cases, hundreds perhaps. (Refer to pathologists reports Google, Avian Flu autopsy studies). In addition to the untold horror of a few untold human to human Transmissions, what would have happened if the avian flu Teamed up with the Swine flu in 2009 and beat the human genetic Defense system to death?

Sequence of how the virus spreads:
Animal / man
Animal infects animal
Animal infects man
Man infects man
Small groups of men to more small groups of men small groups to bigger groups
Bigger groups to pandemic – billions could die in short time

Progression
Bird – to bird
Man eats or cuts bird, (reports in Vietnam, eating birds, or
duck-cooked or not cooked is infectious.)

Man is infected

Man survives or dies

Man infects other men around him

Pandemic-massive deaths up to 50-72% mortality rate for Avian flu.

Cluster

A cluster in CDC report means small groups of people then progressing to multiples of small groups, then to larger multiple groups of larger numbers, first by the 100's then into the 1000's, etc. The CDC has 6 stages described for the Avian Flu. The stages are detailed earlier in the book.

AGRICULTURAL SUGGESTIONS:

Current method of chicken management:

1) Quarantine:
   Mass killing of millions birds chickens China, Romania

2) Creation of a New chicken and or a new egg happened in Britain in November of 2005, cloned a new gene in a chicken which created an immunity to H5N1

A Request to the CDC and FDA, from me.

This testing sequence request was sent to the CDC
December 22nd,2005.

This is my formal request to the CDC, WHO, and NIH to:

Implement a quick and modest double blind study of mice, chickens, pigs and cows etc.,to verify the efficacy of Betadine™ , and its aerosol on animals. Agricultural diet-feeding change and experimental feeding of animals by mixing their food with anise and or licorice, should be ok and may bring about a healthier chicken, etc. Expose the animals to minimal dose of . Iodine in their water and an aerosol of the barnyards and chicken coops and by adding minimal doses of iodine tor their food.

i. Use small amounts of Iodine in the water supply to see if it increases survival rates around infected water or just in general.

ii. And to test as well the efficacy of licorice and anise or other supplements on livestock against H5N1.

iii. A request for CDC to be testing this newly suggested plan on H5N1, and Ebola, Sars and Small pox, especially in the theory of aerosol Betadine™ in homes/buildings.

iv. Test Small pox against PI-Betadine™ especially in the theory of aerosol of buildings homes and cities, regarding biochemical warfare tactics.

v. I believe an aerosol of Betadine™ may actually kill small pox.

vii. **Reinstate vaccination of the world against small pox. It is an eon since small pox vaccinations were given.** *Although unlikely to occur, an Outbreak of small pox is the biggest threat to life on earth.*

Notice:

The CDC has received this Proposal from me December 22nd, 2005.

After test feeding 3 months with licorice and anise, others and Betadine™ and the spraying of the barnyards –chicken coops and the chickens, the CDC can compare survival rates of Betadine™ treated and special test—fed non-fed mice, chickens, pigs and cows with those not treated and the test will show if the Plan had any beneficial effects.

"We are what we eat" is how the saying goes, so why not spend a little more money feed those pigs and chickens and cows something to help them resist infection, and for us to eat more safely.

It might cost $ 5.00 a pound for hamburger, but $ 5.00 a pound is better than no cows at all!

Summary of Tests proposed and justified.

The Japanese have proven effectiveness of PI-Povidone Iodine on Avian influenza A .

1-Select feeding of chickens animals anise and licorice.
2-Spray barnyard and coops with Betadine™ aerosol.
3-Insert select amounts of "Betadine™ into the water.
4-Use Betadine™ topically on the farmer, * HEENT etc.
5-Subject chickens to H5N1 and see if select fed, aerosol-treated are better survivors than the non—fed, and non –sprayed animals.

Results of a simple test like above will possibly show a higher survival rate in the treated, sprayed and fed animals than in the non-treated, and give hope for similar results in the human sector.

Agricultural Habits and the Medical use of Betadine™*

Peroxide—H202 and Betadine™ are reported in the agricultural bulletins on the internet as used by farmers to clean their hands. Protection of farmer using foot ,shoe baths as described and the washing of clothing in veridical agents.

The main concern is to keep farmer from walking the feces etc. into house so the farmers soak their feet in trays with their shoes still on in solutions of virucidal agents.

Please submit you're own ideas to:myself, address at back the book or to the CDC center for disease control, Atlanta Ga.

Summary of a new Human and Animal Plan suggested in this book.

The human plan involves the introduction to idea of Betadine™ for head areas and for masks and for filter systems world-wide. Licorice and anise and other medical nutritional supplements would be added to a routine diet study or pragmatically find a better food substitute. Spend a dollar and save a dollar.

There is no existing medical-chemical-filter-mask system in the world to back-up the deficiency in vaccines except for the systems proposed in this book.

[See the Appendix for design of the mask and filter.]

LAVBAC+FM are letter which **stand for:**
Licorice, Anise, Vitamins, Betadine™Cloud, Filter, Masks.
Human treatment should emphasize the use of
of Betadine™ on all portals of entry of the human body.

Past preventive measures to stop airborne viruses are inefficient and have shortcomings; for example Tamiflu™ is only modestly available and is on backorder and designed only for chicken to man transmission.

The masks are used a lot in the orient but do not protect anyone from anything—they are too porous! The pore size is much bigger than 1u. 1u means one-micron, and it is $1/40^{th}$ the size of the hair on your head!) The Asian and almost all USA masks are psychological at best, not medically efficient at all against Avian Flu. **There are not enough masks or gloves in the USA to service a Pandemic**

BETADINE™ POVIDONE-IODINE

Chemistry and Experimental Uses . . .

Betadine™ Is a 50 yr old antibacterial antifungal antiviral agent, the strongest on earth and is proven and documented Dermatology of 1997 and Laryngoscope 2004. (see references).

The oversight of not using Betadine™ at all on H5N1 appears like running out of house to go to work— literally running by the waiting car in the driveway, and catching a bus, instead of using the car!

We forgot the earlier applications of Povidone-Iodine
completely ignored its original uses.

Betadine is used about well over 50 million times daily across the world, and used to be used on the HEENT body portals.

Betadine™ is the strongest antiviral agent on earth which can be used in every orifice of the body without harm it can and has been put **onto a brain in** neurosurgery and thousands of patient's ears, eyes nose and throat & vagina without harm. Other agents such as **formalin, benzalkonium chloride and ethylene dioxide** are virucidal but cannot be used on or in human—especially Avian entry points . . . namely the HEENT which means, head ears eyes nose throat. Skin test is a couple drops of PVPI on the wrist and see if a Rash appears;

Betadine™ was used to kill HIV on operating room objects, and used to treat certain Herpetic lesions, and is safe in and on human body and has been used for sinusitis and eye and brain surgery.

Chemistry of Betadine™ discussed and referenced

(Betadine™ should not be used for those allergic to shellfish or iodine.)

Betadine™ is organically bound iodine, Povidone is an inert carrier which is neutral, but renders iodine, completely non—corrosive and almost painless. The Incidences of reactions to it are less than 15 cases over a ten (10) year reported study .

Remember that the PVP part of PCPI was a plasma Preservative in Korean War times, and PVP is also now A common meast and poultry food preservative.

Betadine™ has been studied extensively for surgical use; and found to be vrucidal vs. Influenza A & B, adenoviruses, and enveloped and non-enveloped viruses with a better action on the non-enveloped viruses,

The contraindications to Betadine™ use is allergy to iodine or shellfish, but the % percent incidence is less than 15 reports in 10 years of reporting.

Someone should start to use Betadine™ in a more "clever fashion as described below and in this book also to meet the needs of "at risk" humans to airborne viruses. Betadine™ is the most veridical agent on earth which can be put into and onto the human body and is capable I believe of killing Ebola as well.

Nebulizers, Respirators and Emergency Masks

Betadine™ could be used to treat infected viral patients with the older **IPPB or Inhalers and could be used with a small amount of Betadine™** "Mist" to kill the virus in the Upper Respiratory tract.

Someone else has found an eye saving method of using way of using Betadine™ chemical to save a persons eyesight.

Other data:
Data was obtained from the Pharmacology labs of major manufacturers of Betadine™ and faxed directly to me as a courtesy by the pharmacologists at the companies below. They are not contributors to the document, but merely supplied data requested by me at the time.

* Alcon laboratories, Dallas, TX.
• Purdu Fredrick, Stamford, CT,
   The data was faxed directly to me in 2005 upon
   informal request

The following 2 pages were obtained in November of 2005 from the pharmaceutical division of a large manufacturer of Povidone Iodine, along with 36 pages of time studies for activity duration efficacy of the medication.

The point of the following pages is that if Betadine™ is able to destroy all of the organisms listed beyond anything else, and is safe for all uses on a human being, then it would follow that this type of research should be replicated in the CDC,WHO etc; under controlled conditions.

It is senseless not to pursue something that everyone is using in extensive ways with *incredible success in the treatment of eye viruses, herpes ,and even sinusitis in ways that the FDA does not actively support.*

Excellent physicians who trained me, and those whom I have never met swear by the broad uses of this medicine or chemical.

I hope that the studies on pages 83 and 84 will be followed by a lot more pages by the CDC, or at least some chemical which is humanly safe can **help save our lives in a pandemic containment fashion both human, animal and in home and building safety codes. Additional tests from Japan are located in the back of this book And,the study and was done by a similar group 9 years ago.**

Antimicrobial Action of Povidone-Iodine

Dermatology 1997;195(suppl 2):29–35

Inactivation of Human Viruses by Povidone-Iodine in Comparison with Other Antiseptics

Abstract

Inactivation of a range of viruses, such as adeno-, mumps, rota-, polio- (types 1 and 3), coxsackie-, rhino-, herpes simplex, rubella, measles, influenza and human immunodeficiency viruses, by povidone-iodine (PVP-I) and other commercially available antiseptics in Japan was studied in accordance with the standardized protocol in vitro. In these experiments, antiseptics such as PVP-I solution, PVP-I gargle, PVP-I cream, chlorhexidine gluconate, alkyldiaminoethyl-glycine hydrochloride, benzalkonium chloride (BAC) and benzethonium chloride (BEC) were used. PVP-I was effective against all the virus species tested. PVP-I drug products, which were examined in these experiments, inactivated all the viruses within a short period of time. Rubella, measles, mumps viruses and HIV were sensitive to all of the antiseptics, and rotavirus was inactivated by BAC and BEC, while adeno-, polio- and rhinoviruses did not respond to the other antiseptics. PVP-I had a wider virucidal spectrum, covering both enveloped and nonenveloped viruses, than the other commercially available antiseptics.

95

Table 1. Virus strains tested

Virus	Strain	Preparation by
Adenovirus	human adenovirus type 5	Y.Y.
Mumps virus	RW strain	A.Y.
Rotavirus	(RRV) MMU, strain 18006	O.N.
Poliovirus	type 1, Mahoney strain	M.A.
Poliovirus	type 3, Leon strain	M.A.
Coxsackievirus	type B, strain 3	M.A.
Rhinovirus	type 14, strain 1059	I.M.
Herpesvirus	HSV type 1 HF	K.Y.
Rubella virus	M33 strain	A.Y.
Measles virus	Toyoshima strain	A.Y.
Influenza virus	A/Kitakyushu/159/93	O.M.
Human immuno-deficiency virus	type 1	N.Y.

Table 2. Commercially available antiseptics used for experiments

Abbreviations	Main components	Commercial name
PVP-I solution	povidone-iodine	Isodine® Solution
PVP-I gargle	povidone-iodine	Isodine® Gargle
PVP-I cream	povidone-iodine	Isodine® Cream
CHG	chlorhexidine gluconate	Hibitane® Concentrate
AEG	alkyldiaminoethyl glycine hydrochloride	Tego-51®
BAC	benzalkonium chloride	Osvan® Solution
BEC	benzethonium chloride	Hyamine Solution

SECTION VI

DOSES AND PRICES

Types of Betadine™ available types and costs and instructions with prices. 2006 similar to 2009

Topical 10%

    10% preoperative preparation, random wounds,
    $ 8.00—$ 10.00/ 8oz
    $ 17.00/ qt. quart,
    $ 24.00/ gallon Betadine™

Povidone-Iodine which is Betadine™ basically

    10% preoperative is:
    $ 5.00 per pint .

Hair rinse—has been used in the past, years ago.
Not available but doable with proper instructions.

Aerosol 5%

    $ 8.00/ canister

The aerosol can be made by diluting the 10% to 5% And using any household pump container.

Other possibilities are that Povidone-Iodine

- Could theoretically be used for protecting an entire school yard, class room, home, or building.
- And possibly be used for an entire city state, country, via crop-dusting.
- By making us of it in cloud seeding for rainfall.

Ophthalmic 5%

  $ 22.00/ one oz . (See eye case page 73.)

It is a preoperative preparation for eye surgery as well as the sole treatment for adenoviruses of the eye by some eye specialists medical doctors.

Oral gargle-rinse, 0.5% = ½% Was discontinued a few years ago by some companies but can still be bought and can be formed by diluting the 10% solution to ½ % appropriately.

Vaginal—call your pharmacist or the manufacturer.

Nasal—not available, but Betadine™ has been used for years in head and neck surgery and ENT , ear nose and throat surgery and is likely to have strong virucidcal effects used intranasally , or if inhaled moderately. It is used in treatment of acute sinusitis by many doctors—via nasal packing.

The efficacy of Betadine™ and the duration of activity intranasal and intraoral areas must be CDC-studied.

The life span of topical skin application peaks at 90 minutes but then tapers off gradually.

The manufacturer of Betadine™ is Purdue Frederick Company, Stamford, Ct.

The manufacturer of Betadine™ eye drops is the Alcon Company of Dallas, Texas.

Qualities in Huntsville Alabama is another maker of PI.

Another company named Virucidal is supposedly manufacturing Povidone-iodine but, I am not clear on its status.

Betadine™-medical studies, Purdue Frederick 36 pgs. Betadine™ Ophthalmic, Alcon Co.

Time studies Betadine™ Purdue Fredrick

BETADINE MOUTH WASH

Betadine Mouth Wash is still sold as Povidone Iodine
by some companies.

Iodine

Iodine has been used to disinfect water for nearly a century. It has advantages over chlorine in convenience and probably efficacy; many travelers find the taste less offensive as well. It appears safe for short and intermediate length use (3-6 months), but questions remain about its safety in long-term usage. **It should not be used by persons with allergy to iodine, persons with active thyroid disease, or pregnant women**

Betadine™Mouthwash data.

How does it work?

Povidone iodine is an antiseptic. It is a complex of iodine, which kills micro-organisms such as bacteria, fungi, viruses, protozoa and bacterial spores. It can therefore be used to treat infections with these micro-organisms.

Povidone iodine gargle and mouthwash is used to treat infections of the mouth and throat, such as gingivitis (inflammation of the gums) and mouth ulcers. It is also used for oral hygiene, to kill micro-organisms before, during and after dental and oral surgery and hence prevent infections.

What is it used for?

- Infections of the lining of the mouth and throat, such as gingivitis and mouth ulcers
- Oral hygiene before, during and after dental and oral surgery

Warning!

- If symptoms persist for more than 14 days, seek medical or dental advice.
- This medicine should be used to gargle or rinse the mouth for up to 30 seconds. It should not be swallowed.

- Regular use of this medicine should be avoided, particularly in people with thyroid disorders (see below), as prolonged use may lead to iodine being absorbed into the body. Do not use for more than 14 days.

Not to be used in

- Allergy to iodine
- Children under six years of age
- Thyroid disorders such as nodular colloidal goitre, endemic goitre or Hashimoto's thyroiditis (do not use regularly in these conditions)

This medicine should not be used if you are allergic to one or any of its ingredients. Please inform your doctor or pharmacist if you have previously experienced such an allergy.

If you feel you have experienced an allergic reaction, stop using this medicine and inform your doctor or pharmacist immediately.

Pregnancy and Breastfeeding

Certain medicines should not be used during pregnancy or breastfeeding. However, other medicines may be safely used in pregnancy or breastfeeding providing the benefits to the mother outweigh the risks to the unborn baby. Always inform your doctor if you are pregnant or planning a pregnancy, before using any medicine.

- This medicine should not be used regularly when pregnant or breastfeeding. The use of this medicine should be limited to a single treatment session only. Seek medical advice from your doctor.

Side effects

Medicines and their possible side effects can affect individual people in different ways. The following are some of the side effects that are known to be associated with this medicine. Because a side effect is stated here, it does not mean that all people using this medicine will experience that or any side effect.

- Irritation of the lining of the mouth and throat
- Side effects within the body if excessive amounts of iodine are absorbed

The side effects listed above may not include all of the side effects reported by the drug's manufacturer. For more information about any other possible risks associated with this medicine, please read the information provided with the medicine or consult your doctor or pharmacist.

How can this medicine affect other medicines?

People taking lithium therapy should avoid regular use of this medicine.

If iodine is absorbed into the body from this medicine it may interfere with thyroid function tests.

How safe can it get?

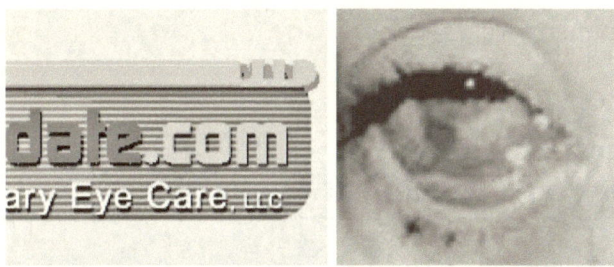

Adenovirus of the Human Eye, treated using Betadine™ ophthalmic solution 5%.

This is a documented case of an airborne virus infection of the human eye—called, Adenovirus of the eyeball. The patient recovered after using ONLY 2 DROPS* of 5% Ophthalmic Betadine ™ Recovery as described was realized within 48 hours, which is very dramatic.

" KILLING THE ADENOVIRUS: *A Medical Cure*

We remain concerned that patients continue to suffer needlessly with adenoviral infection (epidemic keratoconjunctivitis) when a medical cure is readily available. While there is no FDA-approved medicine with such an indication, we discovered a few years ago that use of an FDA-approved "Betadine 5% sterile ophthalmic prep solution" excellently serves this purpose."

This is an example of the safety of managed care of the eye. Other literature shows Betadine's ™ use in the treatment of sinusitis as well as Herpes and prevention of infection of operative instruments from HIV which does in fact encompass the entire upper Respiratory Tract—the chosen sites of entry of H5N1 and all airborne viruses, SARS, Ebola, etc.

Try to expand the forgotten uses of Povidone-Iodine which is Betadine™ to reduce the spread of any and many infections of any magnitude.
The above example supports the misunderstood uses of Betadine™ as I suggest in the text. It has more meaning than a topical antiseptic, it is not a " Kiddie" first-aid toy.

Being safe for the Human Eye opens New Horizons: In a more "clever" fashion consider the use in salt trucks and winterizing the USA against disease of human viruses.

Agricultural Betadine™ Aerosol or Water Treatment Applications

(These are theoretically very doable and cost-effective ideas.)

Agricultural uses . . . of LAVBAC. Licorice, Anise, Vitamins Betadine™Aerosol, Climate Control

Betadine™ is used for farm and human subjects for hand Cleaning after leaving the chicken coops.

Other Possibilities are:
Betadine™ for use on and in head ears eyes nose and throat, even hair and clothing for aerosol control of buildings cities countries and controlled rainfall. **Cloud seeding promoting iodized virucidal rainfall is theoretical but worth a look! It may protect crops as well.**

We should get agricultural grant money for selective test-feeding of animals, domestic and , some for the use of anise seed and licorice which is useful, or a supplement of your choice. Food with additives can be air-dropped for migratory birds anywhere and anytime.

Even C135 Cargo airplane dropping of preselected foods treated with licorice or anise or any antivral supplement is worth a try by putting it in the feeding path of migratory birds or any suspect carrier animal.

Please use anise and licorice under your doctor's advice for a better immune system. (Licorice can cause hypertension, and comes in non-hypertensive forms, called DGL-diglycerinated licorice extract)

Being safe for the Human Eye opens New Horizons:
In a more "clever" fashion consider the use in salt trucks and winterizing the USA against disease of human viruses.

Federal prisons in Florida gave 100's of prisoners Iodine treated water and after 10 years and about 3 penitentiaries No one ill, and only 5 caes of hyperthyroidism occurred! So I2-Iodine is lethal to fish, but it helps people create Thyroid hormone and other stable receptor sites for longevity.

Salt Mine Truck

USING THE SNOW BELT AND SALT TRUCKS
Hypothetical Use of Povidone Iodine.

To Iodinate the USA "snow belt "areas is logical. Salt prophylaxis provides containment of an **Outbreak** of Airborne Viruses, and possibly Biochemical Warfare items.

70% of the USA is in the "Snow-Belt" and using salt trucks loaded with Betadine™ could be allow it to be a national surface antiviral agent.

This will allow the streets and highways to be conductors-of an antiviral agent which is spread over the city homes, Offices and playgrounds which can then vaporize into the atmosphere. After entering the atmosphere, the Betadine is turned into rain and raining Iodine and Betadine onto the surface of the earth.

This can be used for non-viral outbreaks as a prophylaxis against the pollution of storm damage as in Hurricane Katrina.

Try to implement a serious study of the use of Betadine™ for humans and animals now.

SECTION VII

ANARCHY

Whether an outbreak occurs now or in 10 years, now is a good time to educate and enforce mandatory precautions, such as food storage etc. Use this years bird flu for a learning experience before something worse catches us off guard.

Anarchy is a real possibility if the public is not informed as to the reality and implications of what can and likely will happen within a few days to 3 to 10 years, which is widespread death and food shortages loss of communications from a mutation of H5N1 or other viruses.

If the people as a whole are not educated and forced to prepare a bit —a revolution will occur—widespread worldwide disorder and panic, dwarfing Katrina.

As a doctor I found fewer consequences in being honest with a patient . . . than in hiding the truth from them . . . so prepare the people for an outbreak of the following: the Avian Flu, Ebola and Sars . . . even Small Pox— which still is the wildest killer item on the earth!

Prepare people for and with some basics:
Education
First Aid

Some nutritional advice.
What new "Flu" Laws to expect.

Pay people the courtesy of at least saying "maybe you ought to do this" in case, and ahead of time and insist people save food and water and some basic medications to prevent pandemonium and robbery and murders. The assaults on humans by humans in a mere hurricane are nothing compared to a Pandemic death rate and food shortage.

And, in case people start dying we can have a level of educated cooperation and people who will trust and appreciate a solid warning from the government, rather than people being angry from being blind-sided with a death-note restriction at the last minute.

It is wiser to gradually educate people as to what to expect. Rather than to than to hide the "cancer" from the patient just tell him! I always told a patient if he was dying . . . so he could deal with it and make his peace with God and his family and friends.

It is best to ease the world into a "What if?", and "What do I do now?", for what can happen later than to suddenly impose martial law!

Y2K – is a perfect example of the world preparing for a computer failure, saving food, and water; whereas this Avian Flu, SARS business is far worse than any Y2K could ever have been.

An ounce of prevention is the Truth and it will avoid a Government's Nightmare!

Do's and Don'ts for people and governments.
in case of a pandemic in the absence of a vaccine.

The advice here is factual and theoretical and must be done under a doctors advice, and augmented by verification by the CDC, etc.

Do try to have:

- Mandatory storage of non-perishable food items and water, medications, first aid kits
- Communications link to neighbors-not requiring travel
- Medical gloves be provided free or inexpensively.
- Masks 0.1 micron or less preissued now to public
- Pre—issued, Betadine™ masks, be researched and made.
- Chemically coated masks 0.1u and less studied and immediately consider also the
- Aerosol and ventilation filtration with Betadine™ filters in all homes and buildings transglobally
- Give the world some plan besides a vaccine gap.
- Give a protocol, food, clothing, travel, **medical suggestions going far beyond vaccines**!
- Give every citizen a plan they can do and afford.
- For those allergic to iodine, provide special vaccine availability.
- Try to implement studies for Select Feeding of humans and animals some of the ancient herbal treatments such as anise, licorice, pine needle tea, beets, etc.
- And we must make public quickly aware of a protocol Besides waiting out in the cold for a back-ordered vaccine or one that will never be there at all.
- Tell the public what to do and not to do, example the possibility of travel restrictions, permits to drive, etc.
- Reinstate small pox vaccinations and others forgotten for all people under the age of 31.

Small Pox returned after ½ a century to California in 2006 with one isolated caseBut no one under the age of 35 has ever received a Pox Vaccination—so Who has the vial. And what else?

IODINE VALUES vs ALLERGY
No doctor that I have interviewed in Thyroid disease has ever seen an Iodine overdose, and in 31 years maybe has had a handful of cases of iodine induced Hyperthroidism.

Iodine toxicity or allergy:
Is a name without many examples. In 10 years an eyedrop company had only 15 cases of adversity and likely the number of uses in the World was possibly 150,000,000 making this an innocuous agent of help. **It is easier to treat a possible iodine toxicity than it is** to face the fury of a Mutant Killer Virus which is reported to be able to kill 1-3 billion people quickly.

Skin Test For Yourself.
A few drops of Betadine™ on the skin is often the only test needed to test for allergy, a "rash" or "itching" will occur. See your doctor or pharmacist.

Do nots: A Government Must not:

- Fail to tell the public everything now, not later.
- Hesitate to spend an arm and a leg on H5N1,Ebola,
- Leave people to the last minute to go jobless!
- Leave the work place "unsafe" to maintain
- Forget to insure life support systems water heat and electricity functional at the home unit level!
  * Forget to inform all citizens that
- A loss of 1/3 USA work force will occur quickly if a mutation occurs—so plan ahead!
- Betadine™ heating and cooling filters and vaporizers can save a home and building, **I think,it is a real sin to**
- **allow people to have no plan except a vaccine**.
- **Nerves shot?** Well just add this one: a serious pandemic will close down banks, and the stock market.

If All Hell Breaks Loose
be prepared and use Betadine™ and travel restriction which moves may avert annihilation. The simple straight line plan . . . dot to dot is a medical-chemical alternative to vaccine and **deserves attention now!**

If it is safe to do and your doctor or you get skin-tested for PVPI allergy, then people can use a vaporizer in their homes, and vaporize

Betadine™ at a safe level and Discloroation would occur but if you and the neighbors are dying as in 1918 in the USA skip the discoloration for now and I hope that the CDC-will soon test this idea, which may kill all airborne viruses!

It is my suggestion to test the theory of PVPI or the original Betadine™ or a similarly functional chemical be used on people HEENT sites head eyes ears nose throat to **prevent airborne entry to the body.**

In addition plan to use Betadine™ to aerosol a city and thereby be containing and controlling a hideous virus, and reduce the density of the virus and not the people. Literally Iodize buildings, school yards, homes or better yet use some climate control of cities states and major countries, and even entire continents.

Massive storm systems can be used to inoculate infected areas of the countries-even continents can be cloud-seeded to rain iodine onto virally contaminated sectors or areas . . . for example Katrina, Wilma, the Tsunami of last year, which killed 241,000 , and Infected the rest to an unknown extent. If all hell breaks loose—be prepared and use Betadine™ and travel restriction which moves may avert annihilation. The simple straight line plan . . . dot to dot is a medical-chemical Alternative to vaccine.

Vaccinations must be updated.

A well thought out Warning ⇔

The health officials should consider if all Americans under the age of 30 should get a small pox vaccination. Vaccination was discontinued 30 years ago! The USA has 75% of the earth's smallpox vaccine but how can it vaccinate anyone if another virus cripples the delivery system a lot of people will never receive a vaccine for small pox so to me, we seem really naked for a terrorist attack or leak from a lab.

HEPA Safety Factors

Currently the size of H5N1 is listed at about 1 micron = 1u. The minimal range of airplane protective systems to cleanse your travel air have a minimally effective range of .08-1.1—1.3 microns and the plane filters do not block Avian Flu at all.

The problem is the Avian Flu is not in the "optimal-best size" range to be trapped by your airplane! (Refer to the earlier story of lady zero.)

SECTION VIII

NUTRITIONAL

Vitamins traditional suggestions:

Vitamin C
is recognized—for aid in wound healing and resistance to infections. It stimulates production of interferon, a chemical which is often studied in cancer research, which does harm to viruses. If your kidney function is normal, you can take from 100mg—to a 1000mg 3-5 x daily. People under stress take more, and smokers need more.

Minerals.

Selenium is being talked about a lot lately and recommended.

Zinc—seems to be noted a lot in literature; for stress and muscle tone and some antiviral properties.

Copper
Is reported to support antiviral activities.

Med Hypotheses. 2000 Nov;55(5):456-7.

Cysteine, glutathione (GSH) and zinc and copper ions together are effective, natural, intracellular inhibitors of (AIDS) viruses.

Sprietsma JE.

Manganese
Manganese used also with another aerosol was helpful in protecting mice which were infected with Hong Kong Flu. MnSOD was the aerosol used to shield the mice.
American Society of Microbiology 1996.

Magnesium
Is reportedly helpful in some illnesses, heart integrity.

In 2009 verification of a triple mutant swine flu emerged showing Traits of the Avian Bird flu and Poultry and Pigs and Human Respiratory capabilities.

In 2006 this article continues:
To date very few people have been infected with bird flu's. At present there is an outbreak in birds of an extremely pathogenic strain of bird flu in parts of Asia, the H5N1 strain. A few humans have contracted this strain directly from birds. (Australian Text) *[There is a danger in using fish currently due to the Great Lakes Pollution with mercury levels being too high to be fed to pregnant Women> EPA warnings appear on the cans coming out this Fall.]*

L.A.V.B.A.C.-FM

Items included.

Anise
is described as a plant and besides its content in Tamiflu, it has been used for many centuries as a 'healingseed" . . . . although low in vitamin content is highly antiviral and anise seed is distilled to an alcohol and

then to an acid powder called shimick acid and shimick acid is the main ingredient of Tamiflu Vaccine.

Roche™ uses the Chinese Star Anise exclusively. Anise's biochemistry has strange unfamiliar names of chemicals attached to it. Anise is sited as being antiviral, but contains far less vitamin content than licorice. Anise has been used as a dietary fiber supplement and has minimal folic acid and minimal vitamin content. Plain anise alone does not have the nearly the same effect of the Pharmaceutical Tamiflu, but may have some intrinsic value.

- Tamiflu chemistry of is a 6 carbon molecule apparently an acid made from likely an alcohol step from anise powder to alcohol to acid powder.
- Shimic acid is the by product of distilled anise seed star anise of china, apparently the seed is distilled to alcohol and converted to a stable acid powder, to the best of my understanding.
- Out of the capsule, the capsule content has an acid taste and actually bubbles a bit on the tongue.

Licorice
is a member of the "pea family" & grown in Arabic, Mediterranean countries. Is described as having antiviral properties is described by the ancient Egyptians and Greeks for stomach maladies, ulcers.

Strangely the B vitamin content of licorice is very high. The Chemical contents of simple Licorice surpass any Single item I have read about; I intuitively think it is versatile enough to be of great benefit for general health and immunity.

- Paba – Para-amino-benzoic acid,
- Pantothetic acid
- Folic acid
- Vitamin c
- B vitamins
- Others* (see references at end of book.)

- Licorice to me is amazing in its composition and multiplicity of many vitamins from such a simple plant.

Licorice can cause hypertension: so ask your doctor and buy the DG form of licorice.

Nuts-

are extremely helpful. Regular party store seeds are a source of energy, but molybdenum and boron are found in almonds, brazil nuts and the higher forms of nuts.

SECTION IX

EXPANDED USES OF BETADINE

Restrictions, Tests, Normal uses of Betadine™

LAVBAC+FM
HEPA FILTER PROBLEMS
METEOROLOGY-CLIMATE CONTROL
OUTLINE of A NEEED MANDATORY RULES
MASSIVE GROUND INNOCULATION with PI-Salt Trucks

LAVBAC +FM includes the following items.
Licorice, Anise, Vitamins, Betadine™ Aerosol and Cloud seeding, and Crop dusting of cities, and more.

Filters and Masks made from Betadine™

Airplane & Home & Building safety standards

Related to H5N1's size.

H5N1 is 1 micron in size. Almost all planes flying are not "viral" safe nor are they able to filter Ebola, SARS, corona virus or H5N1. Cars, homes, buildings are not the least bit safe from an airborne virus attack.

Refer to earlier National Geographic fictional projections of termination of the human race by a lady called "Lady Zero", traveling from china to London. Dangers to us were noted on TV by National geographic, which sited the following items of concern:

The dangers are those of not eliminating or reducing the air content of H5N1 of not reducing crowds and not restricting air travel primarily from asia to the rest of the World.

All of these will or can lead to oblivion for the human race

It appears to me that an ostrich mentality has developed over the years, especially in USA,no one really believes this will happen or really believe "it won't happen here or if I don't think about it or if I don't see it , then it will go away!

Simply refer to the graphic photos at the beginning of the book of the 1918 plague and history repeats itself viciously.

Over time, depending on the migration of H5N1 and others, worse things will happen in next 3-4-7 years. This is a strong consensus among the majority of virologists.

China October—predicted a work force labor loss of 35% of all people if the H5N1 mutated; consequently how do we live without 1/3 of our labor force or worse—how do we live without imports from China?

Concluding advice:

I would advise that all homes: and public places use this protocol which I call **LAVBAC +FM**

We seriously need to think about constructing some 3M Betadine™ masks, gloves, along with some travel restriction and using the **available DNA-H5N1 testing** kit, which delivers an answer within one hour of flight departure of a land or flight suspected carrier.

As the virus threat worsens by the number of birds infected, and people infected, according to the CDC staging protocol, immediate implementation of the protocols listed should be started. Namely we need LAVBAC+FM, food stamps & work permits, etc., much like the WWII, rationing protocol.

Theoretical Proposed Use is
CLIMATE CONTROL AND CLOUD SEEDING WITH BETADINE™ POVIDONE IODINE.

Imagine if this Hurricane had been medicated with Povidone—Iodine deterring water & sewage infection wherever damage occurred or dumped in the water inlands.

This could be the largest "can" of **spray disinfectant** available on the Earth! But massive or not it makes you think at least now if someone wants to save your life using a water sprinkler you would not be so overwhelmed after reading this.

Iodine can kill the fish population but currently the infections of the Great Lakes are a close 2nd place to Iodine. VHSv – *Viral-Hemorrhagic-Septicemia killed* 100,000, tons TONS of fish about 7-8 years ago in the eastern Lake Michigan area. The mercury is so bad that our Michigan Fish is being labeled **Dangerous** by the EPA for pregnant women. VHSv vanished in 2007 and is back again in Detroit City adjoining waters. Zebra mussels are pouring Alive out of the water faucets in northern Toronto cottage type areas.

So, what I am saying if someone doesn't consider removing some mercury-laced fish and kill a few fish to save Evolution and re-seed the lake after debridement—**there will be no lake to put a fish** in! The lake is an X-file and hidden from us like The UFO's There is a actual situation in Michigan which VHSv *Viral Hemorrhagic-Septicemia* Virus-- country of origin unknown or undisclosed, actually knocked off 100,000 Tons of fresh fish by a leech attaching to side of the fish's head above the eye starting internal and external hemorrhage and death. So if the Mutation capacity of Swine and avian picks up the human adaptive gene- let's call it HAG,of the infected fish, then we have a death bed in the lakes and a death warrant to the the entire planet who takes our water home or visits the lakes for whatever and every ship taking water home for bottling is now spreading an air-waterborne virus sensitive to PVPI only, so if you hear very intelligent People say **'Blowing' the lakes might be the only way to save them,** it is not even talking..about death to the earth, it is about killing the darn virus VHSv and the Mercury and lithium and chloroforms that now make safety impossible without strategic non-lazy moves…treat the lakes like a doctor treats a patient and intelligent ideas will begin.

The lead and the Zebra Mussels outnumber the fish. It might be necessary to disable the life of the lakes one at a time leaving 4 active and one down to restock the dead lake. It is the way nature rebuilds and an aquarium is structured. So never say never to the though of why not take the great lakes down one at a time To get life useable, and why not have a dose of PVPI around in case the bubonic Plague comes back and small pox by aggression kills half of your family in a couple of weeks. **When death comes home to you,** massive medical concepts like this hurricane are friends not fools, and save lives where indicated by real medical judgment, not a biologist reading a slide without a medical team interpreting the data.

Please read this.
This is a Warning of what is happening and never linked as a happening. The carp fish in Michigan used to be less than a foot long not good eating really. Several years ago someone in the Mississippi Delta imported

Indonesian algae To clean up the ocean debris…Suddenly massive mutant Carp emerged. FOUR feet long, and aggressive jumping out of the water hitting people in the head Injuring children in the boats in Michigan. Algae from Indonesia caused this, what do foreign boats have like this. The electric gates massive electroshock underwater fences were broken From the New Orleans Old Miss to Chicago where the Lake Michigan gates broke And now we have 4 foot long Alien Carp hitting kids in the face on the Rouge River Thanks to someone importing algae from Indonesia. Who is supervising what? Who runs the Mississippi.

Last catastrophe:

All of the water from the Big 3 Great lakes Superior Michigan and Huron are running Downhill 2 billion gallons a day due to the Army corp of engineers dredging Operation at the foot of the bridge years ago. Now 2 billion gallons of our water Is going south to Detroit. Ruining the water table down down and down killing more fish Plus another 30-50 million gallons a day taken by foreign ships. But what happens when the VHSv virus runs down hill to meet The Mississippi River near Chicago and the Ocean of the St Lawrence Seaway to the Left or East. If the Zebra mussels of The Great Lakes and the VHSv somehow get whirlpooled into Detroit, by gravity I cannot even guess the outcome.

I want to explain I am a guy who trained in surgery four years and worked ER about 20. The Medical problem is the junction of the Mississippi and the Great Lakes at Chicago is joined with the big blackhole dredged in Detroit, which is joined to the 5 largest bodies of natural water on the earth.

Mississipi + Great Lakes+ St Lawrence Seaway + Joker. The joker spreads infection and inhibits containment of many problems in the Michigan lakes. I need to see the flow charts of where a Lake Superior infections, and Lake Michigan or Lake Huron metal contaminants ends up in the lake patterns of flow.

The Joker is the sinkhole at the Detroit River where water flows backwards from Northern Michigan. It is an oddball and hard to predict, but it seems if there is a contamination in the Lakes it will flow downhill to the rest of the country, but I have not studied or seen where the flow ends up. But this is a possible reverse flow sequence spreading contamination.

In medical terms this is an abcess (big infection) drained by **fistula the St Lawrence Seaway tributaries and originating** from a hole In the Detroit River and the direction it takes carries every infection known to a fisherman and every chemical known to the metal manufacturers west to Chicago and east uphill to NY and out the the Ocean with The boats.

If we don't trash check boats better than the 911 planes, we wont have a lake to jump in And we wont be here to complain about it. This problem is Medical and is reaching a point of no-return. The doctor will come out And tell us They didn't make it…the lakes are gone. 100 policies are working none under any competent board of supervision Take the 20 smartest men on earth bring them hear where 20% of the Earths drinking water comes from.

Katrina—the New Orleans Zoo "Miracle"

Is a stupendous lesson to learn and follow. The workers at the New Orleans Zoo had a plan in place in case of a hurricane, and they saved every animal while everything around them in the zoo was damaged. The city of New Orleans did have plan that they did not use And lost more lives, and the zoo lost none! The New Orleans *Zoo* reopened: November 25th, 2005. The city is still not close to the hygienic conditions of the Zoo as of April 20th,2005.

So a word to the wise: If a zoo can prepare and save animals without one fatality. so should we also have a similar plan like the zoo had. The Zoo Keeper was an absolute artist in the way he was able to save and move elephants and lions and hundreds of animals from a hurricane that killed in the same city, and even keep them sheltered and fed!

Lotto Ticket Vaccine is exactly your chances,maybe less! We are absent of any doable plan at present.

Besides a dependency on a vaccine whose efficacy is totally "lotto" dependent on the mutation rate of the Avian Flu Virus, There is no medical or chemical alternative. There are not enough masks and gloves to save the USA In case of compulsory Pandemic Protocol occurring. It is a dangerous way to live without another choice. Submit to the people of the country, a method to obtain cheap, quickly attainable food, water, vitamins, possibly Betadine™ and related products, gloves, masks (chemical, non-chemical, and also, make-new building & home filter systems of 0.1u (micron) which can function with Povidone Iodine (PI),in order to better protect families and workers. The world needs an outline of "What do I do for the Bird Flu?" It must be precise, and with the best and worse case scenarios elaborated honestly. We should be talking "turkey" to ourselves—wide open—as a world, about the danger of airborne viruses, and doing things we don't like to do, to avoid annihilation.

The anticipated Killer Mutation can leave ½ the earth gone in less than one year. Most experts expect a killer strain in 1-5 years. Special medications and special foods are immediately available and the vaccines are in the distance.

VACCINE can NEVER ,NEVER DEFEAT THE H5N1 AIRBORNE VIRUSES

Fill the " GAP" before Time Runs Out.
Vaccine is like an army without enough soldiers,
Even the eggs to make the vaccine are running out!

The best choice we have is POVIDONE IODINE, called PI or the creation of a new synthetic chemical medication with a similar viracidal spectrum to handle the environment and to take the weight off the vaccine production. Fill the GAP of an understaffed vaccine army with Povidone Iodine or a synthetic or look in the barrel for another one we forgot. *I have infinite respect for immunology, but this time it cannot deal with the numbers if something goes wrong quickly.*

Vaccine cannot deal with the numbers, but the method I propose can offer some help and I can think of no other plan-medical chemical which is this economical or doable.

Care must be offered to iodine allergic folks, in the form of special preference for vaccinations.

So why not try proven Betadine™, and ancient licorice and anise (constituent of Tamiflu) and vitamin C, so at least we are "somewhat protected" instead of standing in the cold waiting for a vaccine shot?

A Plan for All Seasons and All Countries

MUST-HAVE TO-DO '*NOW or NEVER*' PLANS.
The following are *normal public health* measures As simple as a bar of soap compared to the No-plan currently recommended.

The current plan of vaccine as a remedy is like using an umbrella to stop a bullet!

Vaccines cannot get the gun out of the holster fast enough . . . it is a plan of high stakes at high risk in all respects. I ask that Betadine™ be strongly reconsidered in its every aspect, even using a vaporizer in homes and offices if and before the H5N1 or Sars comes back to haunt us. World wide recommended protocols of this book are summarized in word LAVBAC which is an economoical,easy bet for global safety.

Old and New Protocols

1.
LAVBAC+FM:
Licorice Anise Vitamins Betadine Aersol Cloud Seeding Filters and Masks and Gloves.

Food items are discussed in prior sections and the Betadine™ air and surface treatment of homes, public places as well as the placement of Povidone Iodine vaporizers for home, office major locations will help immensely.

2.
Nutritional:

Institute the use of vitamins and minerals to humans, pets and farm animals and insist on imperative water treatment of farms homes and buildings with Betadine™ and similar medications.

Preventive medication and *food supplementation on all farms,* inspection of all produce, grain and livestock.

4.
Travel limitation:

City governments decide who needs to travel and if it is helpful to a pending viral outbreak that certain folks be on the road *then issue permits.*

5.
Termination of Public Events

Including public gatherings of football, basketball, baseball and all spectator sports and the closing of bars and restaurants if this flu breaks loose.

6.
Mandatory DNA testing

Preflight testing and everyday testing of all people

- DNA testing is available for H5N1 infected passengers
- And is also useable for daily people.

7.
Heat scanners

Are being installed at many airports work places for fever detection. Viral heat sensors are—currently in use at airports USA to detect infected or "on –purpose" intruders carrying a lethal virus, such as H5N1 or Ebola carried in from the Eastern areas. The sensors will detect people having a fever, thereby allowing the triage of suspects for further investigation.

8.
Around the Corner stands Blind-sided Danger.
A straw to break the camels back.

Consider the danger of seeing who is at your front door while the thief is at the back door.

In gearing up to fight the Avain flu we have found out:

a. that there is no chance for further vaccine production after 450 million doses
b. we are running out of eggs
c. we are unprepared for anything new or old.

The survivor of the present danger may not be the most powerful, but the least affected or the best protected or best—prepared to sadly be 1 of 8 living survivors of that entity able to inflict a viral attack on a group which has no defense or vaccine delivery system, so violent an attack that only 1 of every 8 homes will have a living person in them. In 1918 30,000,000->100,000,000 died supposedly from H1N1 and I agree but never knew from school and supposedly because they knew less than us . . . . we know nothing more than then of protective help except an if come vaccine and PVPI which is what I am about.

Again if you cannot deliver older shots from the vaccine bank because you are Shut Down by a Pandemic, then you are politically,and medically crippled.

If an enemy attacks us during a vaccine shortage with any of the older viruses we will be defenseless.

How will we defend ourselves, if we

- a.) cannot unpack old vaccine
- b.) cannot make a new one
- c.) one for lack of manpower
- d.) two for lack of eggs.

With Our Backs To The Wall and all around us *are dying including all livestock and fish, then perhaps we can venture into the Twilight Zone of Chemical-Medical Management of Pandemics and Or their prevention – allow me to offer a cure when all around you are dying or before it happens, no one has a plan human safe For air and water but I am working on it.*

We should Lean On A Known Established Medical-Chemical which is able to Kill Airborne Viruses or Invent New Ones!

Where and when needed pull out the stops and kill some viruses and be willing to lose some normal flora to: Pandemic stage 6 Means there will be less human race so play with big toys Not little toys or you will die.

BETADINE™ POVIDONE-IODINE

"BETA-RAIN"
short form for Betadine™ rainfall

Make it rain and snow Betadine™

If the theory I propose is incorrect then try to improve the older patent or replace it and deal in the real world of designing a **medical –chemical patent which is antiviral and human-safe.**

No chemical made on earth is as universally safe to humans and as broad spectrum in germ killing as Povidone Iodine-Betadine ™

The inventor of the new H5N1 vaccine has said that there will never be enough vaccine to help all of us and that there are not enough eggs to make vaccine and new way to make Vaccine might be needed. The point is if Vaccine can only save 1/12th of the human Race and the other chemicals are not human safe, the PI-Povidone plan is strangely the "Only Show In Town".

Betadine™ and other chemical medical agents-should be tested and considered for aerosol of buildings and cloud-seeding large areas of cities, states and countries and continents.

Road-treatment with Povidone Iodine

Theoretically,
A Truckload could save our lives

Salt mines and salt trucks and Betadine™

The use of local salt mines to add Betadine™ to the salt-trucks loads can utilize the snow to afford massive coverage and literally— tracking of antiviral footprints . . . into homes businesses and social places, and to promote the evaporation of ground H20 into the atmosphere, initiating an iodine-type-antiviral rainfall, from which I coined the name " Beta-Rain".

The sewage and fresh water monitoring plants

might benefit from a bit of 'old fashioned iodine'. We need to keep a careful eye on sewage and fresh water supplies reservoirs that may house viruses. Mandatory minimal requirements for food, water in houses, must be enforced to avoid anarchy. Insist that every and all vaccinations be updated and for sure I am asking that all small pox vaccinations be resumed. See the prior section on Anarchy.

Vaccinations must be updated.

Education and research grants should be offered for all students entering research. Genetic alternation of chicken's farm animals for resistance seems helpful

SECTION X

DOSES AND COSTS OF VITAMINS, LAVBAC

LICORICE ANISE ZINC VIT.C BEEPOLLEN BETADINE POVIDONE I

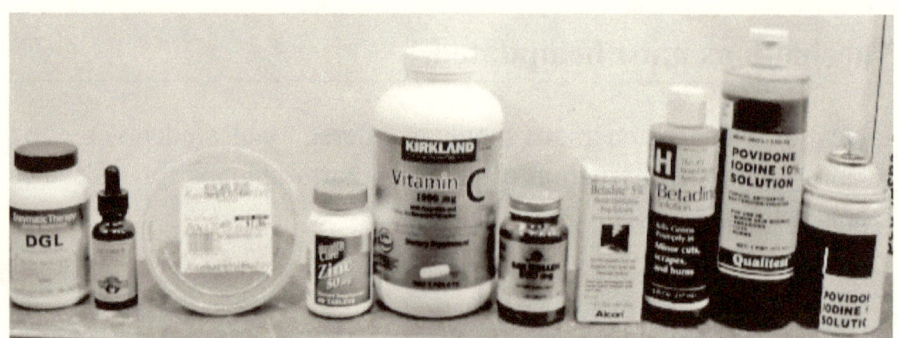

LICORICE ANISE ZINC VIT.C BPOLLEN BETADINE AEROSOL CLIMATE CONTROL

Photo of Sample Nutritional protocols – what do they cost?

Table of Items: Photos Description Doses and Costs

The doses listed here are for standard illnesses, ulcer colds, bronchitis, but there are no recommended doses for a killer virus, H5N1 etc. So ask your doctor or nutritionist for their suggestions.

Anise

Dollars and doses—how much to take.

Anise.
Anise tastes like licorice. It can range from $ 4.00b non-organic to 16.00/lb organic. To my mind anise is anise,the non-organic is expensive. Mediterranean and arabic cultures have used it as a pastry sweet-seasoning-topping for over 2000 years.

Anise Costs.

Type	$/lb	Doses
Turkish	- 5.00/ lb seed >5.00/lb powder	2-3 tablespoons 3x / day can mix with cold tea/ pop/etc
Chinese	>$ 10-20/lb	

Licorice Root

Licorice comes as a tablet or liquid extract-primarily from licorice root and it can cause hypertension. So it is available to hypertensive patients as a "DG" Tablet/Liquid which means –that it is " a dilycerinated form". It is apparently Acceptable for use in patients with HTN—high BP-Blood pressure. It is actually a member of the pea family and not exactly a plant.

Licorice continued.
Standard health food store licorice tabs
$ 8.00/ 100 tablets to take 2 tablets before each meal

Licorice liquid extract.
$ 10.00 ounce and to take a dropper full before each meal.

Canadians Harvested Thousands Of Pine Trees for Vaccine Content

Pine Tree

Pine Needles

have been used for hundreds of years as well as birch bark for their antiviral properties.

Recently thousands of trees have been harvested in Canada in February of 2006 for their "Tamiflu"—Shimic acid content found in the pine needles of the Northern Pine Tree.

The cost is dependent on your own ingenuity and location. Tea is made by boiling the needles. Caution see a doctor Familiar with this item.

Bee Pollen

has old-fashioned claims for help for health
Cost is $ 4.00/bottle of 150 tablets.

Vitamin C

100 mg / #100= $ 3.00	3x /day
500 mg / #100 $ 5.00	3x/ day
1000 mg/ 100-500 $ 10—$ 15-$ 20.	Etc 3x/day

When you take Vitamin C be sure your kidney function is normal.

Zinc
50 mg $ 3.00 1-2 tabs daily

Children doses
Recommendations are not known at this writing & the use of Betadine™ is an established Fact & theory , some of the book is factually useable, and the Theoretical parts are labeled as such. And you must consult your pediatrician

Selenium
Is highly recommended for its antiviral properties.

Minerals:

Copper
*Med Hypotheses. 2000 Nov;55(5):456-7.*Cysteine, glutathione (GSH) and zinc and copper ions together are effective, natural, intracellular inhibitors of (AIDS) viruses.
Sprietsma JE.

Manganese
Manganese used also with another aerosol was helpful in protecting mice infected with Hong Kong Flu. MnSOD was the aerosol used.
American Society of Microbiology 1996.

Magnesium
has to do with cardiac function and has reported antiviral properties and other medical illness benefits.

Porphyrins-
A large number of natural and synthetic porphyrins of diverse chemical compositions and characteristics can be isolated from nature or synthesized in the laboratory. Antimicrobial and antiviral activities of porphyrins are based on their ability to catalyse peroxidase and oxidase reactions.

Filters and Masks.

See the Appendix for specialized discussion of a newly proposed mask and filter system.

The newest 3M mask and very few others provide 0.1u blockage of viruses. A good single respirator mask is $ 1.59 to $ 2.50 each depending on the dealer. Medically safe filters for airconditioning,and heating and cooling units do not exist at the present time anywhere on earth. See Appendix for a new method of deviralizing the air at the home and office

Costs of Betadine™ and is also sold as Povidone Iodine.

Betadine™ in Breastmilk true true

CNN broadcasting last June of 2008 that Betdine is being used by some in the breast pump for breast-feeding of HIV infants resulting in a reduced mortality record.

Betadine™ Topical—10% solution $ 25.00/gallon

Povidone Iodine Topical sold in one pint containers $ 5.00/container x 4= $ 20.00/gallon

Dilutions of the above were/are used in nasal packing for sinusits by some doctors.

Povidone Aerosol USA/Japan about $ 5.00 small can.

Betadine™ hand aerosol Pump $ 8.00-$12.00 plus per 16oz.

Betadine 5% Opthalmic eye drops pre-operative medication: One ounce costs $ 22.oo

Betadine™Gargle or mouthwash is sold by Povidone Iodine companies worldwide currently and sold as a ½% solution.

Vaginal Desuche for infection may still be available in some countries.

SECTION XI

FIRST AID

Suggestion to Hospitals and Paramedic Teams: A Theory worth testing immediately. Betadine™ could be used to treat infected viral patients with the older **IPPB or Inhalers and could be used with a small amount of Betadine™** "Mist" to kill the virus in the Upper Respiratory tract. Or it could used prophylactically in the field in a respirator system to kill incoming pathogens, even biochemical pathogens may be stopped.

2 Types of emergency kits:

1. Routine daily ER kit (non viral-kit)
2. Viral bacterial chemical ER kit

1. Routine Daily Emergency Kit:

Breathing
Bleeding and
Shock

Most all emergencies are in these categories, and the list is just a basic mini-list, see back of book for A detailed list.

Stethoscope, blood pressure cuff, Glucometer
Thermometer-or strips, urine test strips. Oxymeter.

Breathing: equipment

- oxygen-masks,tubing,tanks for elderly
- Bronchodilators-alupent, etc
- Decongestants-Sudafed Actifed etc.
- Plastic oral airways

Bleeding: equipment

- tourniquets
- 4x4 gauze pads, bandages, band aids etc.
- being up to date on immunizations,especially Tetanus
- Abds and ace wraps
- Suction machine

Shock: equipment

- vasopressors
- Adrenalin , bee sting kits*(have adrenalin)
- Benadryl injectable and pills
- IV hospital solutions
- IV tubing etc.

CPR: equipment is complex expensive & requires licensure to operate
safely, defibrillators etc.

Other basics:

Medications "old-fashioned things we used to use"
ASA-aspirin/ Tylenol
Baking soda for rashes
Bleach for cleansing dirty areas***

(caution on vapors)

KI—is potassium iodide and recommended by the USAF for nuclear war protection of the thyroid gland; consult your doctor for how to get and use. It is also the historical agent of choice for cloud-seeding To induce rainfall artificially.

2. Viral—Biological Kit.

The contents include the same as the above routine E.R. items and a **few extras** Suggestions here are for a viral outbreak and not just infection of chicken.

Basic Survival Items.

1) shelter
2) clothing
2) food & water
3) waste disposal(personal and trash)
4) heat and lighting
5) transportation
6) entertainment

If an outbreak occurs, 1/3 of the work force is lost in 2 months time, and more work force shortly after the first few months *.A Shut-Down of Global Economy will possibly occur due to an Asian Outbreak, paralyzing the China East-West trade balance.* Consequently, there may be no power, and no food, no water for 3—6-months if some survive the" mutation" or "outbreak". There will possibly be no heat, no air-conditioning, and contamination of sewage and toilets resulting in gross infection city and country wide.

These are the items you need to have on hand. and can be used if a Pandemic occurs or not.

Problem and Answer

1) **Shelter**
   homes/offices
   lock it against intruders seeking food etc.
   secure it against airborne viruses and even duct tape your windows.

sewage disposal:

   trash bags for feces
   bottles for urine disposal
   tissue paper,tampons
   decontamination of air, floor

- The following is a suggestion as to how to deviralize or decontaminate your home:
- Aerosol of vaporization of the Povidone Iodine.
  Which is nont-tested on H5N1 yet but the
- **vaporization or aerosol killing of viruses and bacteria including HIV, is done in hospitals by using Betadine™**, PI. The spray is used to disinfect surfaces primarily but has a good potential ' air-decontamination' effect.
- Aerosol of homes, buildings and iodine toxicity has not been researched by CDC yet but will likely soon be done and will likely save your life if you are not allergic to iodine.
- Betadine™ is Povidone Iodine. A simple few drops on the skin will tell you if you're allergic or not.
- Gargle of the Povidone was sold in the past, but it should not be swallowed. It takes enormous quantities of ingested Betadine™ in an adult to cause toxicity but do not swallow it on purpose.

Gargle is useable.

- Overdoses are actually treated by ingesting large amounts of bread which the starch binds the Iodine molecule. (Home test: Put a few drops of Povidone-I
- on paper,Kleenex or bread and it will turn blue.)
- or use a wet noodle which is pure starch for an iodine test.
- See the section on Betadine™ doses, modes for HEENT, head-ears-eyes-nose and throat uses and even vaginal uses.

Betadine™ is dark brown-" dark brown iodine color" and it will stain clothes furniture and drapes –everything you love or like, but a pandemic will kill you and the discoloration will not matter. There is a vitamin store which sells a clear substance which is labeled Organic Iodine, but it is KI—potassium iodide and it does **not have** the same viracidal effect as Betadine™.

Theoretically, using a vaporizer with a modest solution of Betadine™ mist and aerosol in the room will likely kill or reduce the amount of every known germ, virus, bacteria in the house, office, or building,but this must be tested by federal health folks first before using this idea.

The CDC must recommend or test this for safety. This would involve a lab test of H5N1 with Betadine™

2) Clothing.*caution coming into your home from outside.

We must have lots of all-seasons and transitional clothing because an outbreak can last more than 6 months and you have to survive on your own resources. The clothing: must be *carefully washed* if it was used outside the house or at work or from the outdoor Cleaning Agents to be used could be recommended by the USA Government if they are available.

3) Masks

Must be worn out of the house or office they must be 0.1 u = one micron pore size tightly face fit.

- Masks do not kill viruses or anything, the virus is possibly blocked on the surface of the mask so keep you fingers off the cover of the mask and don't rub the mask then your eyes or food or you may die by accidentally touching your mouth or nose or eyes; this is fact –this is confirmed

- That if H5N1 , Ebola or small pox are on the surfaces of your mask, and enter any head orifice or lungs you can become infected.
- Gloves and masks should be worn while out of the house, or to handle any delivered items.

4) Food and Water Supplies.

Me I love pizza, but consider this:

"Where did the ingredients come from and who has the driver been in contact with recently?

"Where is the chicken from?"

Food and Water Requirements.

My estimate is to have about 50 gallons of water per person stored in the house or office

Food: about *50-100 pounds **of rice per person***

Include dried mild and dried cereals etc.

The reason for suggesting **rice** (almost non-perishable for 1-2 years) is that current refrigeration systems may not be working. Other dried goods compatible with powdered mild or water are also good to have around. Salt, sugar ,dry powdered milk,cereals etc.

5) Heat and Air Supply

Portable space heaters (preferable Betadine™ coated filter systems for viracidal effects furnace and a filter changes often (possibly Betadine™ coated. for outgoing air decontamination.) The items of filters and heating and cooling ideas have to be CDC tested before using and are theoretical at this point.

6) Transportation

You may not be able to use the car
Permits to travel on certain days will likely be needed.
A serious thought for home and car:
Keep gloves and wipes in the car,

> "bird feces" may actually kill you, if the bird is infected with a mutant strain".

7) Entertainment and Religious services

- If an outbreak occurs,
  There will be no pubic functions
  No football, no hockey, no baseball etc
  No religious services
  No funerals

- So find some literature to read
  Internet services may be abandoned for lack of personnel to service the accounts.

- News people will not be doing " live coverage

- Marital law will likely exist nationwide.
- Be prepared to help the country, stock up
- Food and water etc like a Y2K.

8) Communications will be Impaired.

If we are quarantined by the federal government in the event of a pandemic outbreak we will be forced to live, work, study and communicate through telephones and the internet. If this should happen we'll all be communicating via video email live webcasting and video instant messaging.

Why not? Put a drop in the Bucket of Life?

We throw condoms and everything else anti-environmental into our water so *why not* save some lives if a "flu" goes wild, and put a drop of Iodine in the Bucket —of the earth like the "good old days".

BET ON A TRAVELED MEDICINE

Vaccine is a "maybe deliverable" on time commodity, but the Betadine™-Povidone Iodine is a "here now" medication and has been here 50 years", so if the CDC approves its continued use on humans, expanding its uses, Betadine™, Povidone-Iodine will offer a tangible hope of survival to all of 6.2 billion inhabitants of the earth, and the estimated 50-100 billion animals on the Earth! At least produce a better Povidone type system. Afterall it is about 50 yrs since it was designed!

These recommendations will need CDC and agricultural studies, but this is *not** off by much. Get together and talk it over and write your leaders! This is worth a town-meeting for a start

These recommendations will need CDC and agricultural studies, but this is *not** off by much. Get together and talk it over and write your leaders! This is worth a town-meeting for a start.

Please talk the book over and approach someone to research the Povidone-Iodine or Chemical –Synthetic—Antimicrobial-Medicine Theory and to invent other similar items.

F. J. Sawaya M.D.

SECTION XII

SUMMARY

SYNOPSIS—NEW PROPOSED SYSTEMS

This is an excellent summary of the entire book in outline form.
Format for prevention of avian flu H5N1

To the reader:
The following is divided animal and human methods of curing viral disease prior and current, and new proposed suggested systems.

Use an Influenza A screening test, example QDEL test to screen patients, and a *medication like Betadine*™ to contain an outbreakand *normal first aid* supplies for confinement food power etc. QDEL tests for Influenza A and the Avian flu is a subtype of influenza A. And the Swine flu called H1N1 is also a subtype of Influenza A

(a)	(b)
Current System	**New Proposed System**

Diagnosis, Treatment, Prevention, Research
Environmental & Miscellaneous.

1) Animal   2) Man

I.)
CURRENT SYSTEMS
used for animals and men in the prevention of the Avian Flu.
Diagnosis—treatment—prevention—research, environmental control and miscellaneous.

A) Animal-Agricultural.

Diagnostic:

Most all diagnosis is done on the appearance of the bird, feathers, weight etc.

Preventive:

Sick birds are killed by the millions, leading to an egg shortage worldwide. The department of agriculture and farmers have some strict and interesting protocols for the cleaning of their shoes going from the barn to the house and use a footbath to clean their shoes. Betadine™ and peroxide H202 are used often.

Treatment of birds

Infected birds are killed by the millions, and currently in China,after killing 43 million chickens to stop an outbreak is Also injecting a vaccine which is being used which is inferior to normally used vaccines, and can **cause mutations** of dangerous proportions as the DNA of the chicken and vaccine and a common flu cohabitate*

Research

Agricultural research is far more profound and extensive and sophisticated than human data. Autopsy reports are complete exact and extensive.

B.) Human: 2006 and 2009 now August.

Since 2006 when I wrote this data,comes the Swine flu The Swine flu is a respiratory aggressor into the Human Lungs And so is the Avian flu and now both directly attack people By skipping the turkey the duck and the pig. It's a way of getting the pig to fly, the airborne virus inside a pig or bird can by air enter us and kill us.

Current
System prevention diagnosis treatment research environmental and miscellaneous.

Diagnostic:
DNA testing is available now for plane travelers
and can reveal presence or absence of H5N1 within
one hour of boarding a flight.

Heat sensors—are now hidden in most airports
to detect terrorist entry with Ebola or H5N1 travelers who will be having a fever, and pulled aside for examination and questioning.

Symptoms-
Upper respiratory, fever, temp, malaise, rapid pneumonic Decomposition.

Treatment:
Is primarily via Tamiflu vaccine,which has saved some lives in Asia who were actively infected. But often if the human is actively ill, the care is respiratory and is supportive only and carries a mortality rate of 52-70%. (world-wide figures.)

Vaccine available:

The following data is from November 2005.

But remember: there is no vaccine available for human to human transmission until the "outbreak-mutation occurs and then there is a 6 month gap until the first single dose of a useable vaccine can be given to one person, and during which type upwards of 2 billion people can die. [National Geographic, TV special nov.2005]

Actual vaccine available to USA in 2006 was 4.3 million And went to multi-millions as winter approached. The 2009 Swine flu vaccine is *not the right one to stop* human to human transmissions.

2006 Avian statistics:

There are only 4.3 million doses of Tamilfu vaccine USA Roche is backordered on Tamiflu. However some state There are one billion doses available with a shelf life of only one year. Toward the end of 2006 **Hungarians** have new highly effective vaccine—the maximum production capacity is 2million/mos or 240 million doses/year.

So even if the Hungarians license the patent to 20 licenses, then 20 licenses will make 40 million doses /mos. But remember that 6 billion doses are needed not counting animals. Argentina making a Tamiflu this Spring of 2006.

Prevention:

All the masks used in Asia China, and Japan are non-effective and will not begin to block the 1u size H5N1 virus particles! The only totally safe mask I understand is a 3M O.1 u (micron) mask.

A recommendation of the world health leaders on Meet the press November of 2005, television show indicated that *not frequenting public places would be a wise move for people to follow, that is avoiding large crowds.*

Quarantine and or confinement

will be necessary; should an outbreak occur. All non-medical and non-military personnel with be confined to their homes, sick, infected or not and work from home, and shop based on their street address, odd or even numbers and be forced to shop with mask and gloves on at the stores they frequent.

II.)
NEW SYSTEMS proposed
Diagnosis, treatment, prevention, research, Environmental control, & miscellaneous.

Agricultural & Human Combined Techniques.

LAVBAC-FM as discussed
Licorice, Anise, Vitamins, Betadine, Aerosol, Climate Control
Filter and Masks

Diagnostic:
Earliest possible detection can limit an outbreak. DNA testing must be used in all sectors, not just airplanes.

Heat sensors must be in all sectors and not just airports, Much like the system in England of TV spotters video tapes running and really reducing the crime there.

Skin testing
Explore other skin tests for H5N1, Ebola, and Small pox A suggestion of mine is to try cobalt light reflected on skin treated skin with perhaps antigen/ antibody diluents for a quick skin scan.

Treatment:
In the USA there is a shortage of 6 million hospital beds

In case of an outbreak of proportions of 25%

And this is serious! With the Swine flu there is actually going to be A shortage of 25,000 ventilators but that's a low figure

Try my guess of 150,000 ventilators and beds if a pandemic starts but the total number of treatable people will be less Than 10-20% of 300 million USA population. Methods of treatment in hospital must have a new wrinkle, breakthrough.

Treatment: Prophylactic
Nutritional.

Test the effects of: new foods for our animals

Example:
Licorice & anise to be fed to animals, or some change in diet of expensive proportions to create a sturdier useable food supply, resistant to infection both domestic and wild life & jungle animals.

Prevention: Stop Asian or all Foreign Birds
Block imports & exports of all birds
Reduce all travel from China & Southeast Asia

DNA testing mandatory.
Heat sensor placements everywhere.

Proposed New Ideas:
Prevention/ treatment, diagnosis, R&D and **environmental control.**

New prevention measures-medical:
Cloud seeding fish-safe areas with Betadine™
Will really alter the viral environment

The only contraindication to using Betadine™ Is an allergy to iodine and shellfish, but even then The chance of adversity is slim, none and zero: Reports from manufacturers of Betadine™ products

And a 10 year computer report input study showed only 15 cases with reported side-effects from Betadine™ in a ten year span. It is used over 25 million times daily, so the risk of adversity is nominal.

Animals/ Part 2 of
Human and Animal Protocols Suggested

Betadine™ aerosol

^

Barnyards, clothing   **farms, homes offices**

Testing needed to be done, a method of doing it.

1) Test aerosol spray chicken coups and barnyards with Betadine™.
2) Insert tiny doses of Iodine to the water supply.
3) Test feed ½ of selected animals
    Licorice and anise, and/or other supplements of study
    And choices of the CDC & departments of agriculture.
4) Select feed the animals a special diet, possibly Licorice, anise pellets
5) Pre-treat the animals with iodine-laced water and/or aerosol.
6) Then infect "select-fed" "sprayed" animals with H5N1
7) Infect "normal-fed". Non-sprayed" animals with the virus.
8) Results: will show effects of anise licorice Betadine™

It is possible that animals and people using the aerosol of Betadine™, and minor water treatment, with a better diet will have a lower mortality rate than the untreated animals.

9) The estimated time to test theory of all three items is about 6-8-12 weeks.

Nutritional suggestions for animals.

Air drop licorice,anise seed or similar antiviral foods in the migratory birds paths . . . use a C135 airtransport plane to do it.

Nutritional suggestions for farmer's diet

Try putting anise and licorice into the farmer's diet along with vitamins and minerals in doses high enough to effect a change improving general immunity along with vitamins and minerals in dose high enough to effect a change.

Vaccinate the farmers and not the chickens since vaccinating chickens breeds hideous Frankenstein mutations of deadlier H5N1 viruses. With the exception of Betadine™ all other effective anti-H5N1 agents are not safe for human use and may not be as effective as Betadine™ which is human-safe.

Treatments "old and new include those which are unsafe for humans are:

> Benzalkonoum chloride
> Ethylene dioxide
> Formalin.
> Bleach
> Semi-safe agents are:
> H202 – peroxide & Betadine™ are used in farmer
> shoe wash as practiced currently-dept agriculture studies.

Betadine™ meteorological uses.

Theoretical *
Crop-dusting of local areas, major cities or counties can be done. The contraindication of an allergy to iodine in human use is to be tested as Betadine does not bring about identical effects as plain Iodine, but caution is needed. The spectrum of PVPI is actually 30% greater than plain Iodine.

If the Iodine-Air-Content were tested by the, CDC or DOA—Department of Agriculture, the Iodized air may not reach harmful levels for the allergic or non-allergic people!.

Cloud seeding with Betadine™

Modified cloud seeding over selected areas with be beneficial, and not harmful in modest doses to the animals, farmers or consumers.

Human body respiratory entry points

The use of Betadine on the farmer and his farm and his animals, even iodine in the water may effect a change in the world for the better. Allergies to iodine, are minimal and would require select doses of vaccine perhaps. But in my own 30 years of using Betadine™ and from the computer reporting center; the incidence of adverse reactions to Betadine™ are slim. You can skin test yourself with a tiny or few drops rubbed on the skin, a rash will appear if you're sensitive.

Farmers Upper Respiratory Care:

HEENT:

HEENT is the abbreviation used by doctors for the following

Head
Ears
Eyes
Nose
Throat

Betadine™-Povidone iodine:
Is available in ophthalmic, hair, oral (was sold as a mouthwash until a few years ago and it is available as room spray and was sold as a vaginal douche.

Betadine™ literally covers every portal of entry & is safe for all human use and possible animals. In the respiratory tract: every portal of entry is safely protected.

Tools are labeled "abcdefgh" and are listedl showing you that Betadine™ covers all parameters of the standard medical protocols used for years.

Summary of treatment-agricultural.

Select antiviral vaccine for farmers*

Select antiviral agents for animals

Clone a new egg make a new chicken

Use Betadine™ for HEENT protection

Use Betadine™ aerosol for coups & use it for barnyards,

Possibly place iodine into water

Prior animal test feeding joined with human care will reduce the viral spread.

NATIONAL RULES for an OUTBREAK
That will be IMPLEMENTED and the
LAWS YOU WILL SEE

Environmental conduct rules

Closing of all public entertainment facilities

Confinement to home, work from home Is likely to be a government mandate.

Shopping will be allowed by your Street address, odd or even numbers And then only occasionally, not up to the public.

You will not be allowed to shop or enter A store without gloves and masks and then only on preselected days of the month based on your street address Being odd or even.

Martial law will likely be imposed.
There are not in 2009 a in 2006 enough gloves for more than 10% of World and not more than 10% of mask protection Of even the incorrect mask and 2% coverage of the o.1u Micron mask for Avian flu . . .

If we are quarantined by the federal government in the event of a pandemic outbreak we will be forced to live, work, study and communicate through telephones and the internet. If this should happen we'll all be communicating via video email live webcasting and video instant messaging.

Sawaya Proposal asking CDC to test all items ASAP

Test LAVBAC+FM

L Licorice
A Anise
V Vitamins
B Betadine™ Povidone-Iodine
A Aerosol Povidone Iodine
C Cloud seeding and climate control, Povidone.
F-Filter
M-Mask

Test a variation of man-agricultural diet change coupled with aerosol Betadine™ mist, Betadine™ heating and cooling filters, also, it can be used in/on many parts of the body.

Human protective measures.
Prevention tools A-H.
See the Betadine™—rain theory.

Nutritional
Licorice
Anise
Vitamin C
Zinc acetate discs for mouth
Selenium
Copper
Magnesium

(doses/costs are listed in another section)

Medical.

New proposal for treatment

Betadine™-is Povidone—Iodine

Betadine™ has been used safely for 45 years in all upper respiratory entry points in the human practice of medicine.

HEENT is head ears eyes nose and throat
Uri—is upper respiratory infection

Betadine™ is available & safe for human use.

HEENT is an abbreviation used by doctors for and describes:

-Head, even a hair wash
-Ears
-Eyes
-Nose
-Throat-topical & vaginal

MANUFACTURERS-SOURCES FOR BETADINE™

Purdue Frederick co. Of Stamford, Ct. makes all items except the eye drops, which Are made by Alcon labs, Dallas, TX.

Japan, makes all and then more than the American versions

And many other American Companies make the Povidone Iodine which is Betadine™ chemically.

TYPES OF PVPI or BETADINE™

General Purpose Betadine is 10% solution.
Eyes: . Betadine™ 5% ophthalmic comes from Alcon, Dallas, TX.
Throat wash—is sold by some companies
Vaginal douche—was used in past
Topical scrub pre op surgical scrub
Aerosol pump— available USA, Japan, etc.

Theoretical Medical Alteration of the Environment

Since H5N1 is an airborne virus, It is fair to assume that creation of an "antiviral environment" and even atmosphere will save many lives.

Betadine™ has been used for many years from Japan to the USA in aerosol form in operating rooms to decontaminate the air; consequently, it will help to create an environment similar to an operating room can be formed to turn the home or office or locality into a nearly "sterile" environment.

DEVIRALIZE THE LAND and
MAKE IT RAIN MEDICINE

Theoretical Situation:

Iodize the Salt Trucks in winter.

If Betadine™ is used in salt trucks and then allowed to evaporate vertically into the atmosphere, it will possibly create an anti-viral rain which protects the surface we live on like a blanket! Betadine™ rain can be intensely effective for massive ground cover of a city a state a country and even a continent; think of the reach of the 5 hurricanes this past season 2005. The ecological effect of iodine is Likely negligible compared to an absent planet, but **the ecology and ozone layer problems have to be answered.**

KI—crystals

Potassium iodide crystals have been used for years as the traditional agent of cloud seeding to produce all artificial rain on earth and no one seems the worse for it! It is the agent of choice which will protect the thyroid gland from nuclear Radiation from a mistake or a nuclear warhead bomb.

I think that using Betadine™ in salt trucks in the snow belts will affect a broad antiviral evaporation cloud and antiviral rain, and track the agent all over the cities and homes.

The fact is, about 70% of USA is in the snow belt and a lot of the 70% use salt natural or artificial.

This 70% includes 250,000,000 cars in the snow belt and can create a lot of "vaccinator vehicles." . I estimate the cars have the ability to inoculate the surface area of three quarters of the USA!

Massive ground and air cover is effective possible only with this method

In dry climates or dry or non snow seasons, perhaps the old-fashioned crop dusting plane can be dragged from the hanger and get flying again over infected areas.

Building and home modification of filter systems can provide shelter from the virus. The theory works well in the winter months.

Betadine™ Snow/Rain in a Nutshell

Combining Snow, Rainfall and Filter Systems & Medical-Chemical-Masks, and Crop-Dusting will possibly make a much Safer World all year round!

200 –300 million cars tracking "Betadine™ slush" all over the countryside will make a great blanket of antiviral "Lysol" and have some virucidal effect.

The how much to use and how much the effect, is for the powers to test estimate and determine. But, certainly It will limit the colony count of airborne viruses.

Although H5N1 lives in cells-animal and human; it is an "air-traveler" and subject to the proposed theory.

I believe if enough Betadine were dumped in Only Selected areas in Cambodia or African Reservoirs which can never be found accurately that Ebola could me minimized or eradicated with minimal Damage to the wildlife and jungles.

Satellite Tracking of Ebola Reservoirs

You can even try to " **Isotope Tag" the Betadine™ to Satellite Track its path** in fish (if need be)and animals in the dense rain forest –jungles!

Nutshell Summary

Snow –salt trucks –Betadine™-Snowfall-antiviral surface cover—Betadine Cloud-Seeding, Swamp and sector seeding* for Ebola.,Crop-dusting, Betadine Masks and HEENT applications.

Kill the Ebola reservoir if located.

This would be a bonus option available to the theory of Iodine in the water, Ebola must have a non-human host which I am certain is subject to fatal effects of Betadine™ on any organism.

Bonus option possible:
Inoculate the Cambodian and African swamps on a chance of getting the *reservoir" knocked out.

(dose control will not alter environment, possibly)

Using Betadine™ to seed the salt trucks can cause enough ground air concentration that it will make a nice brown blanket to cover us from the deadly viruses around the corner.

As the water on the ground rises, evaporates and becomes rain, it will actually create an antiviral cloud cover whose "mld" (minimum lethal dose-yet to be determined) is controllable and tolerable if moderated by the CDC and the department of agriculture. The concentration of iodine in the air for this method has yet to be determined.

A new humidifier
Picture 100 square miles of surface rain or snow evaporating and coming back down to earth like a giant blanket and protecting us from airborne viruses.

Betadine™ can actually make us a "controlled environment" and will be most likely be totally effective against the H5N1 virus or any other airborne virus, or bacteria.

Crop-dusting of large areas

Using a plane or making an aerosol of large proportions for infected suspect areas will work and can include entire cities, counties in theory and safely.

Even swamp lands of Africa can be ridded of Ebola-the most hideous virus on earth, by dumping Betadine™ into the swamp lands and crop-dusting, *plane-seeding the area clouds to rain "iodine" onto the Ebola virus and rain forest if needed.

Storm center –Hurricane Inoculation

Had the world tested Betadine™ prior to the 5 huge hurricanes and even tsunamis that trashed Asia, we could have *stopped some of the infection of water and land* with a few plane loads of KI(potassium-iodide) and Betadine™, A medicated hurricane,if possible could have innoculated the Caribbean, Southern United States and Parts of Central America in one dose by using a few plane loads of Povidone-Iodine and KI –postassium iodide crystals.

A medicated storm system will be safe:

- If research allows such massive moves and
- If it is safe for the ozone layer by cloud seeding under 30,000 feet levels.

We could have stopped or reduced a "theoretical outbreak" from ever happening.

We could have prevented somewhat, the biological sewer that occurred in New Orleans, because the water would have had a reasonable content of iodine and prevented cholera, killed some HIV, stopped a lot of bacteria and viral growth from occurring.

Lastly one flush of a toilet in china hospitalized 122 people who nearly died, because the sars virus apparently can move through the rubber or metal seals of the plumbing in the toilet systems in Chinese or other hotels! This actually happened on the 15th floor of a China hotel.

If SARS had hit the Twin Towers, it would have
been a mortality rate of 72% plus meaning about 35,000 dead or critically ill people would have occurred.

So why not have Betadine™ Tidy Bowl, if we want to survive and not join the 100 million dead from 1918 where there were no planes, or trains and barely a car! Our great **grandparents used Iodine for purification of water for the last one-hundred years.** Betadine™ is he strongest humanly safe antiviral agent on the planet Earth, verified and uncontested.

Consult studies were obtained by me from to obtain from the pharmaceutical division of Purdu Frederick co., Stamford, CT.,or Alcon labs, Dallas, TX.

You also can request information perhaps.

Thank you,

Frederick James Sawaya M.D.

This Plan was and is intended to offer something medical/nutritional/and meteorological to save us from a "Viral-Outbreak" in the absence of the backordered and new vaccines, and a dwindling egg supply.

> **TOOLS – to fix the virus etc.**

Routine medical methods used for many years
Medical tools used in standard public health protocols

A) **Direct killing of host**
B) **Direct killing of virus**
C) **Diet anise licorice etc**
D) Masking
E) Medication
F) Quarantine – sequestering of the infected
G) Work restrictions all planet, ban export import
H) Driving, flying all travel restriction except emergency services

Betadine™ or Povidone-iodine is available via

Products of Purdue Frederick Company, CT.
Qualitest Co., Huntsville Alabama, etc. &
Alcon Co, Dallas TX (eye-drops).
And it is available other places and world-wide
Use for about 50 years! Of success.

Betadine™-is available to protect every entry point of a virus or bacteria, even a chemical, Including the treatment of the human eyeball.

The use of Betadine against airborne viruses Is proven in studies to be effective against enveloped and non-enveloped viruses,moreso against the non-enveloped, H5N1 is enveloped.

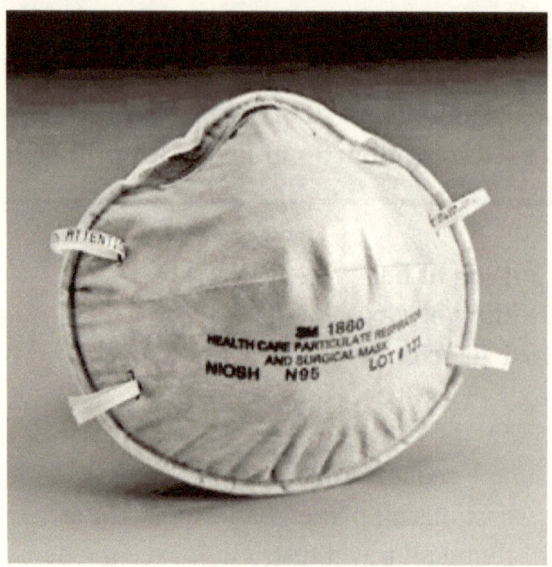

Sample Photo of a 3M Mask,

Current surgical masks were originally designed to protect the patient not the doctor. A tuberculin mask is maybe the safest for use against the viruses.

Our provisional patent describes a special mask, NOT the new 3M Mask which blocks H5N1 but that mask implemented to block like the 3M and Kill the virus.

A mask similar to the above surgical mask has been designed by the 3M company to "Block, Trap" H5N1 and other airborne viruses.

Use a sample 3M mask 0.lu which has oxygen pores and blocks anything over 0.1u 1/10th of a micron.

Betadine ™was effective against enveloped and non enveloped viruses, more so on the non-enveloped.

H5N1 is and "Enveloped" virus, but I feel would
succumb to Betadine easily if Tested.

To the best of my knowledge 30 years as a medical doctor,no infection has stood up to or against Betadine™ or Povidone Iodine and survived. F.J.S. M.D.

New Japanese studies from April of 2006 are now available to prove that Betadine kills H5N1, and we should check if it will kill Ebola or SARS.

It is due to the study from Dermatology 1997,furnished by the pharmacology division at Purdue Fredrick that we feel that Betadine™if used to its full extent-topical to aerosol—that it is very likely kill all airborne viruses in 2006 even H5N1 and SARS, maybe even Ebola. Further research should be done to increase and expand its uses apart from its labeled use. The activity of and duration of Betadine™ effectiveness can be chemically augmented perhaps.

There is supporting evidence in 1997 to support that H5N1 will die from contact with Betadine TM and happily the same source

Supplied the PVPI kills Avian Flu data in April of 2006!

The April 2006 article of the Japanese verified the ability of Povidone Iodine to kill avian influenza A viruses. This makes a strong case For this provisional mask and filter patent to be supported.

Don't take life for granted and have a second angle on this killer—which infected 28% of the USA in 1918 and killed unknown millions of citizens.

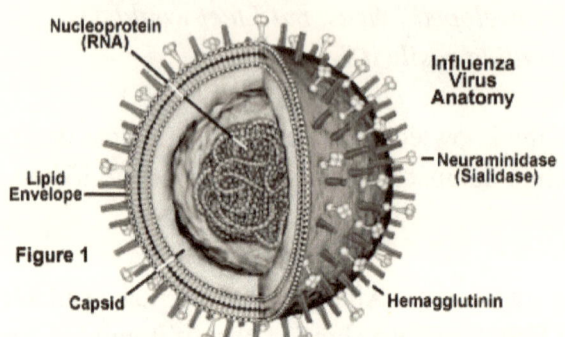

Nucleoprotein (RNA)

Influenza Virus Anatomy

Neuraminidase (Sialidase)

Lipid Envelope

Figure 1

Capsid

Hemagglutinin

A Typical Virus Structure

This nasty critter will travel 10-20 feet with a sneeze or cough and penetrate every mask on earth like a 22 caliber bullet from a rifle shot. The Avian virus will penetrate over 90% of all masks except the new 3M-0.1u mask. After penetrating all preexisting masks except the best, it will have a 72% chance of killing you within 3-14days time even with medical care.

The catch "22" is that not even the new 3M mask can completely protect you; wiping the mask and touching your eye or dishware can infect you, but killing the trapped virus eliminates the chance 100%.

Proof of the theory of Betadine™Povidone Iodine.

This is the beginning of proof of concept that a Betadine™ or Povidone Iodine treatment plan is needed for use on the human HEENT areas,and that the mask,filters,aerosol, crop dusting and cloudseeding concepts as discussed here, have a good scientific chance to save our lives.

JAPAN APRIL OF 2006.

This section is from PubMed, April of 2006. It relates findings of Japan about Povidone Iodine on avian Influenza A. PubMed.gov is the publication.

The results indicate that PVP-I products have virucidal activity against avian influenza A viruses,therefoe PVP-I products are useful In prevention and control of human infection by Avian influenza A viruses"

Outbreak of highly pathogenic avian influenza in Japan and anti-influenza virus activity of povidone-iodine products.

<IMAGE>

SECTION XIII

JAPANESE STUDY OF AVIAN FLU AND POVIDONE IODINE

Outbreak of highly pathogenic avian influenza in Japan and anti-influenza virus activity of povidone-iodine products.

Dermatology. 2006;212 Suppl 1:115-8.
Ito H, Ito T, Hikida M, Yashiro J, Otsuka A, Kida H, Otsuki K.
Department of Veterinary Public Health, Faculty of Agriculture, Tottori University, Tottori 680-8553, Japan.

OBJECTIVES: On January 12, 2004, an outbreak of highly pathogenic avian influenza, caused by the H5N1 strain, occurred in a one-layer flock in Yamaguchi Prefecture, Japan. It had been 79 years since the last outbreak of avian influenza was confirmed in Japan. By February, 3 additional outbreaks had occurred (1 in Oita Prefecture and 2 in Kyoto Prefecture). Influenza viruses are enveloped viruses and are relatively sensitive to inactivation by lipid solvents, such as detergents. Infectivity is also rapidly destroyed by ether, sodium hypochlorite, povidone-iodine (PVP-I), peracetic acid and alcohol. However, these antiviral effects were only tested against human influenza A viruses. In the present study, the antiviral activity of PVP-I products against H5, H7 and

H9 avian influenza A viruses, which had recently been transmitted to humans, were investigated. METHODS: The in vitro antiviral activity of PVP-I products (2% PVP-I solution, 0.5% PVP-I scrub, 0.23% PVP-I gargle, 0.23% PVP-I throat spray and 2% PVP-I solution for animals) against avian influenza A viruses [a highly pathogenic avian influenza virus, A/crow/Kyoto/T2/04 (H5N1; 10(6.5) EID(50)/0.1 ml), and 3 low pathogenic avian influenza A viruses, A/whistling swan/ Shimane/499/838 (H5N3; 10(4.8)

EID(50)/0.1 ml), A/whistling swan/Shimane/42/80 (H7N7; 10(5.5) EID(50)/0.1 ml) and A/duck/Hokkaido/26/99 (H9N2; 10(4.8)

EID(50)/0.1 ml)] were investigated using embryonated hen's eggs. RESULTS/DISCUSSION: Viral infectious titers were reduced to levels below the detection limits by incubation for only 10 s with the PVP-I products used in this study. These results indicate that PVP-I products have virucidal activity against avian influenza A viruses. Therefore, the PVP-I products are useful in the prevention and control of human infection by avian influenza A viruses.

These results indicate that PVP-I products have virucidal activity against avian influenza A viruses. Therefore, the PVP-I products are useful in the prevention and control of human infection by avian influenza A viruses.

PMID: 16490988 [PubMed—in process]

REFERENCES

Inactivation of Airborne Viruses by use 1997.195(suppl2)20-35, furnished by Purdu Frederick Co. of Povidone Iodine

The Journal of Dermatology

Viracidal Effects of Povidone Iodine Tested (Japan) on Avian flu successfully.

Dermatology, April 2006.

Time-Activity Studies Betadine™

Purdu Frederick Co.

Povidone Iodine Spray

Shndong.Jewin Pharm Co.

Anise

Health Touch on Line.

Licorice

www.ayvitamins database

Review Article, Avian Influenza

Indian Association of Microbiology (excellent overview of H5N1 , 2004)

Iodine and the Ozone layer

Google-Ozone Layer data

H5N1 is Developing Resistant Genetic Mutants, 2005.

Google,Mutations of H5N1

Betadine™ 5% Ophthalmic,

Alcon Laboratories, Fort Worth TX.

Betadine™ — Purdue Fredrick Co. Stamford CT.

Iodide — *www.life-extension-drugs.com/iodidekj*

Anise — Health Club On line *www.healthtouch.com*

Time Studies on duration of activity of Betadine™ — Journal of Hospital Infection (1995) 29,9-18.

Answers to Questions about Betadine™ — Internet Access.

Chemical Composition of Chinese Star Anise — JAMA vol.291. No.5., February 4. 2004.Journal of American Medical Association

Betadine™ 5% Ophthalmic Solution — Drugs.com, a product of Alcon Labs. TX.

Reported Antiviral Properties of Essential Oils — Google.

H5N1 is developing Tamiflu Resistant Genetic Mutations. — Google

St Lawrence Seaway Data, Flordia — state prison meidal records.

Sharp's Plasma cluster Ions Deactivates H5N1 — Google

Sulfa*Derm (Volcanic Ash)Promoting Skin function — Google

H5N1 News & Resources about Avian FluExtenive — Google

Bibliography of broad-based and Multiple topics. — Google

Iodide, Iodine Damages to the Ozone Layer	Google
Avian Flu In Turkey? October 08.2005	Google
Latest In Human Nutrition, April-June 2005.	Michael Gerber M.D. about 70 pages.
WHO, World Health Organization,	Daily update on reported cases of H5N1 and others, via Google. Internet.
Deadliest Viruses of the Earth	National Geographic, Television November 2005 and February 2006.
Stories of Southeast Asian Cases of Avian Flu	CNN Television
MSNBC abstracts Internet April/May 2006	MSNBC Internet
The QDEL Test	Life Science Weekyly, 2006
General and clinical information	My Personal Medical Colleagues from 1965-2006, Immunologist, Physicians.
	John E.Magielski M.D.,Univeristy of Michigan
	Herbert Roth,Pediatrician, Michigan
	Karl O. Bandlien, Surgeon, Michigan.

ABOUT THE AUTHOR

Frederick J. Sawaya M.D.
University of Michigan Medical School 1969
General Surgery Training 4 years
Emergency Room Physician 20 years, House Calls 5 years.

F. J. Sawaya M.D.
Authors Background

Catholic Central H.S. 1961—John Carroll University 1965
University of Michigan Medical School 1969
General Surgery 4 years—Emergency Room and Urgent Care
27years—Research in ENT 1970, Patents and Product Designs 31 years.
Originator of the Widely used "Sharp-Trap" Biodisposable Box.

"With the world's attention directed towards the real threat of a pandemic outbreak of the avian flu, it is disturbing that the "solutions " offered to us by our scientific community can only offer to protect a small fraction of the world's population. It is therefore very encouraging that someone is willing to step outside the box of "traditional solutions" to find alternative solutions. Dr. Sawaya's research on the use of Povidone Iodine, which is a widely available and inexpensive product, is very exciting since it indicates that this may represent **a real alternative solution with universal application and immediate availability**."

Karl O. Bandlien MD, Chief of Staff, Detroit Hope Hospital.